Microprocessor Architectures
RISC, CISC and DSP

By the same author

VMEbus: a practical companion

Newnes MAC User's Pocket Book

Newnes UNIX™ Pocket Book

Newnes Upgrading your PC Pocket Book

Newnes Upgrading your MAC Pocket Book

Effective PC networking

PowerPC: a practical companion

In preparation

The PC reference book

The PowerPC programming pocket book

All books published by Butterworth-Heinemann

About the author:

Through his work with Motorola Semiconductors, the author has been involved in the design and development of microprocessor-based systems since 1982. These designs have included VMEbus systems, microcontrollers, IBM PCs, Apple Macintoshes, and both CISC and RISC-based multiprocessor systems, while using operating systems as varied as MS-DOS, UNIX, Macintosh OS and real time kernels.

An avid user of computer systems, he has had over 60 articles and papers published in the electronics press, as well as several books.

Microprocessor Architectures
RISC, CISC and DSP
Second edition

Steve Heath

Newnes
An Imprint of Butterworth-Heinemann Ltd
Linacre House, Jordan Hill, Oxford OX2 8DP

A member of the Reed Elsevier plc group

OXFORD   LONDON   BOSTON
MUNICH   NEW DELHI   SINGAPORE   SYDNEY
TOKYO   TORONTO   WELLINGTON

First published 1995
Reprinted 1995

British Library Cataloguing in Publication Data
A catalogue record for this books is available
from the British Library

ISBN 0 7506 2303 9

Typeset by *Steve Heath*
Printed and bound in Great Britain by
Clays Ltd, St Ives plc

Contents

Appendices

Preface

' Why are there all these different processor architectures and what do they all mean ?'

'Which processor will I use? and how should I choose it?'

There has been an unparalleled introduction of new processor architectures in recent years which has widened the choice available for designs, but also caused much confusion with the claims and counterclaims. This has resulted in questions concerning the need for several different processor types. The struggle for supremacy between complex instruction set computer architectures and those of reduced instruction set computer purveyors and the advent of powerful digital signal processors has pulled the humble microprocessor into the realm of computer architectures, where the total system is the key to understanding and successful implementations. The days of separate hardware and software engineering are now numbered because of this close interaction between hardware and software. The effect of one decision now has definite repercussions with so many aspects throughout the design.

Given the task of selecting an architecture or design approach, both engineers and managers now require a knowledge of the whole system and an explanation of the design trade-offs and their effects. This is information that rarely appears within data sheets or user manuals. This book fills that knowledge gap by closely examining the developments of Motorola's CISC, RISC and DSP processors and describing the typical system configurations and engineering trade-offs that are made. Section 1 provides a primer and history of the three basic microprocessor architectures. Section 2 describes the ways in which the architectures react with the system. Chapter 6 covers memory designs, memory management and cache memories. Chapter 7 examines interrupt and exception handling and the effect on real-time applications. Chapter 8 examines basic multiprocessing ideas. Chapter 9 gives some applications ideas which show how certain characteristics can be exploited. Section 3 looks at some more commercial aspects. Chapter 10 covers semiconductor technology and what it will bring, while Chapter 11 examines the changing design cycle and its implications for the design process and commercial success. Chapter 12 looks at future processor generations and Chapter 13 describes the criteria that should be examined when selecting a processor. The appendices include further information on benchmarking and binary compatibility standards.

Although these comments are as relevant today as they were four years ago when the original edition of this book was published, much has happened within the industry. Since then the PowerPC architecture has appeared and RISC appears to be mounting a serious challenge to the supremacy of CISC. So important are these issues that new material on the PowerPC has been added and a whole chapter devoted to understanding the RISC challenge, as well as the more normal updating of material.

The examples have been based on Motorola's microprocessor families, but many of the system considerations are applicable to other processors with similar architectural characteristics. To emphasize this point, this edition has additional information and comparisons to other designs and in Appendix B, an overview of the other major processors that are available such as the Intel 80x86 architecture and DEC Alpha.

The material is based on several years of involvement with users who have gone through the decision making process and who have frequently asked the questions at the beginning of this preface. To those of you that asked and inspired the original and requested this edition, I thank you.

The application note on high integrity MC68020 and MC68030 designs, the descriptions of the MC68040 processor and silicon technology in Chapters 9, 10 and 12 are based on articles that I have written for Electronic Product Design. Their permission to reprint the material is gratefully acknowledged.

In addition, I would like to say thank you to several of my colleagues at Motorola: to Pat McAndrew who has guided and educated me through DSP — I look forward to working with you again — and to John Letham, Ian Ferguson and Mike Inglis for their support over the years! Special thanks must again go to Sue Carter for yet more editing, intelligent criticism and coffee when I need it.

Steve Heath

Acknowledgements

The following trademarks mentioned within the text are acknowledged:

MC6800, MC6809, MC6801, MC68000, MC68020, MC68030, MC68040, MC68332, MC68302 MC68851, MC68881, MC68882, MC68008, MC68HC000, MC68HC001, DSP56000, DSP56001, DSP96000, MC88100, MC88110, MPC601, MPC603, MPC604, and MC88200 are all trademarks of Motorola,Inc.

PowerPC is a trademark of IBM.

UNIX is the trademark of AT&T.
VRTX is the trademark of Ready Systems, Inc.
OS-9 is the trademark of Microware.
pDOS is the trademark of Eyring Research.
PDP-11, VAX 11/780 and DEC are the trademarks of Digital Equipment Corporation.
iAPX8086, iAPX80286, iAPX80386, iAPX80486 and Pentium are trademarks of Intel Corporation.

1   Complex instruction set computers

8 bit microprocessors: the precursors of CISC

Ask what the definition of a CISC (complex instruction set computer) processor is and often the reply will be based around subjective comments like 'a processor that has an over complex and inefficient instruction set', or even simply 'an MC68000 or 8086'. The problem with these definitions is that they fail to account for facts that more CISC microprocessors are used than any other architecture, they provide more direct hardware support for the software developer than any other architecture and are often more suitable for applications than either RISC (reduced instruction set computer) or DSP (digital signal processor) alternatives. A CISC processor is better described as a mature design where software compatibility and help for software are the overriding goals.

Many of today's CISC processors owe their architectural philosophy to the early 8 bit microprocessors either as a foundation or as an example of how not to design a high performance microprocessor. It is worthwhile reviewing these early designs to act as an introduction and a backdrop to further discussions.

A microprocessor can simply be described as a data processor: information is obtained, modified according to a predetermined set of instructions and either stored or passed to other elements within the system. Each instruction follows a simple set path of fetching an instruction, decoding and acting on it, fetching data from external memory as necessary and so on.

In detail, such events consist of several stages based around a single or a succession of memory accesses. The basic functional blocks within a typical 8 bit microprocessor are depicted in the diagram. The stages needed to execute an instruction to load an accumulator with a data byte are shown in boldface type. Each stage is explained below.

Start of memory cycle 1.

Stage 1:   Here the current program counter address is used to fetch the next instruction from system memory. This address is 16 bits in size.

Stage 2:   The 8 bit instruction is placed on to the data
                bus and is loaded into the data routing
                logic from where it is passed to the instruc-
                tion decode logic.

Stage 3:   The instruction decoder examines the
                instruction, determines what operation is
                required and how large the op code is. In
                this example, the op code requires the data
                value stored in the next byte location to be
                loaded into an internal register. This uses
                several control signals to start the next data
                fetch and route the data successfully.

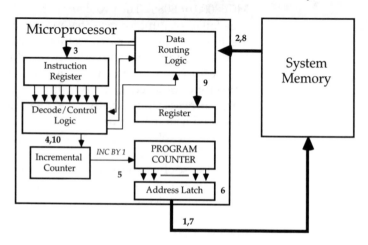

*Stages involved with
instruction execution*

Stage 4:   An incremental counter increases the
                program counter value by one so it points
                to the next location.

Stage 5:   The new program counter value is loaded
                into the address latch, ready for the start of
                a new memory cycle.

                End of memory cycle 1.

Stage 6:   The latched 16 bit address is placed on the
                external address bus.

Stage 7:   The next external memory cycle is started.

Stage 8:   The 8 bit data is placed onto the data bus
                by the system memory and is latched into
                the data routing logic. This logic has been
                told by the instruction decoder to route this
                data through to the register.

Stage 9:   The data is written into the register.

Stage 10:   The instruction decode logic is told of the successful data transfer, the counter incremented by one, and the next address stored in the program counter ready for the next memory cycle.

End of memory cycle 2.

This succession of events is simply repeated to create the processor flow.

8 bit microprocessor register models

The programmer has a very simple register model for this type of processor. The model for the Motorola MC6800 8 bit processor is shown as an example. It has two 8 bit accumulators used for storing data and performing arithmetic operations. The program counter is 16 bits in size and two further 16 bit registers are provided for stack manipulations and address indexing.

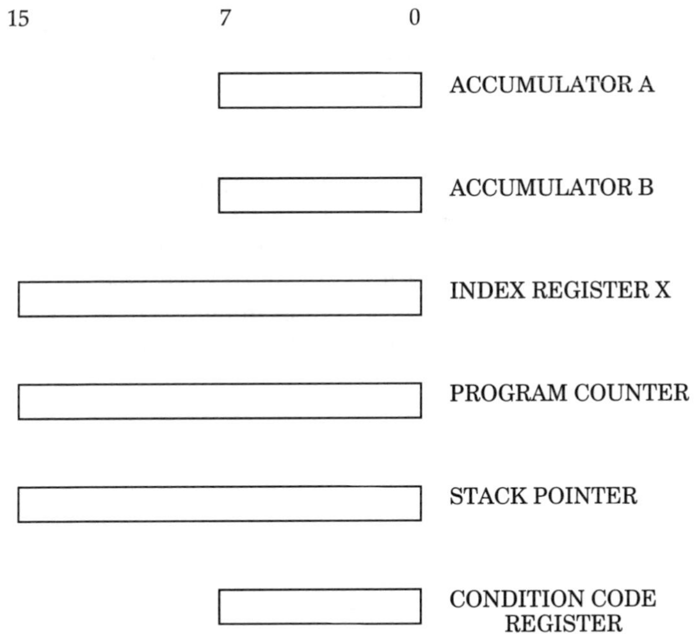

15   7   0

ACCUMULATOR A

ACCUMULATOR B

INDEX REGISTER X

PROGRAM COUNTER

STACK POINTER

CONDITION CODE REGISTER

The MC6800 programmers model

On first inspection, the model seems quite primitive and not capable of providing the basis of a computer system. There do not seem to be enough registers to hold data, let alone manipulate it! What is often forgotten is that many of the instructions, such as logical operations, can operate on direct memory using the index register to act as pointer. This removes the need to bring data into the processor at the expense of extra memory cycles.

The stack pointer provides additional storage for the programmer: it is used to store data like return addresses for subroutine calls and provides additional variable storage using a PUSH/POP mechanism. Data is PUSHed onto the stack to store it, and POPed off to retrieve it. Providing the programmer can track where the data resides in these stack frames, it offers a good replacement for the missing registers.

Restrictions

An 8 bit value can provide an unsigned resolution of only 256 bits, which makes it unsuitable for applications such as financial, arithmetic, high precision servo control systems, etc. The obvious solution is to increase the data size to 16 bits. This would give a resolution of 65,536 — an obvious improvement. This may be acceptable for a control system but is still not good enough for an accounting program, where a 32 bit data value may have to be defined to provide sufficient integer range. There is no difficulty with storing 8, 16, 32 or even 64 bits in external memory, even though this require multiple bus accesses.

However, due to the register model, data larger than 8 bits cannot use the standard arithmetic instructions applicable to 8 bit data stored in the accumulator. This means that even a simple 16 bit addition or multiplication has to be carried out as a series of instructions using the 8 bit model. This reduces the overall efficiency of the architecture.

The code example is a routine for performing a simple 16 bit multiplication. It takes two unsigned 16 bit numbers and produces a 16 bit product. If the product is larger than 16 bits, only the least significant 16 bits are retained. The first eight or so instructions simply create a temporary storage area on the stack for the multiplicand, multiplier, return address and loop counter. Compared to internal register storage, storing data in stack frames is not as efficient due the increased external memory access.

Accessing external data consumes machine cycles which could be used to process data. Without suitable registers and the 16 bit wide accumulator, all this information must be stored externally on the stack. The algorithm used simply performs a succession of arithmetic shifts on each half of the multiplicand stored in the A and B accumulators. Once this is complete, the 16 bit result is split between the two accumulators and the temporary storage cleared off the stack. The operation takes at least 29 instructions to perform with the actual execution time totally dependant on the the values being multiplied together. For comparison, an MC68000 can perform the same feat with a single instruction!

```
MULT16      LDX    #5              CLEAR WORKING REGISTERS
            CLR    A
LP1         STA    A    U-1,X
            DEX
            BNE          LP1
            LDX    #16             INITIAL SHIFT COUNTER
LP2   LDA   A      Y+1             GET Y(LSBIT)
            AND    A    #1
            TAB                    SAVE Y(LSBIT) IN ACCB
            EOR    A    FF         CHECK TO SEE IF YOU ADD
            BEQ    SHIFT           OR SUBTRACT
            TST    B
            BEQ          ADD
            LDA    A    U+1
            LDA    B    U
            SUB    A    XX+1
            SBC    B    XX
            STA    A    U+1
            STA    B    U
            BRA    SHIFT           NOW GOTO SHIFT ROUTINE
ADD         LDA    A    U+1
            LDA    B    U
            ADD    A    XX+1
            ADC    B    XX
            STA    A    U+1
            STA    B    U
SHIFT       CLR          FF        SHIFT ROUTINE
            ROR          Y
            ROR          Y+1
            ROL          FF
            ASR          U
            ROR          U+1
            ROR          U+2
            ROR          U+3
            DEX
            BNE          LP2
            RTS                    FINISH SUBROUTINE
            END
```

M6800 code for a 16 bit by 16 bit multiply

A simple 16 bit digit multiply is a common requirement and yet it requires a complete subroutine to perform it even crudely. Why not provide a 16 bit MUL instruction?

This obvious solution is hindered by the effect of 8 bit data size on processor instructions. First, a 16 bit multiply requires extra registers to cope with the increased data size. These registers require instructions to load them, manipulate data etc. Each new instruction requires a unique op code to define the operation, which register(s) are involved, any implicit data and addressing modes. Unfortunately, an 8 bit instruction width only allows a maximum of 255 different op codes, from which all these combinations must be encoded.

To overcome this, many instruction sets use additional bytes (called operands) to contain additional data which cannot be encoded within the op code. A load accumulator LDA instruction for the MC6800 consists of an instruction plus 1 or 2 bytes containing either implicit data or some form of memory addressing. The penalty of additional machine cycles and slower execution is the resulting performance as depicted and described earlier.

Examination of an op code list for the MC6800 shows another drawback: some instructions only operate on one special register. The advantage is that the register does not require encoding which, considering the 8 bit 255 op code limit, was beneficial. This highly non-orthogonal nature either forces a lot of data movement to and from the registers prior to execution or forces direct memory manipulation with its slower access. Code examination indicates that up to 40% of all instructions executed with such architectures simply move data from one location to another without any processing.

Addressing memory

When the first 8 bit microprocessors appeared during the middle to late 1970s, memory was expensive and only available in very small sizes: 256 bytes up to 1 kilobyte. Applications were small, partly due to their implementation in assembler rather than a high level language, and therefore the addressing range of 64 kilobytes offered by the 16 bit address seemed extraordinarily large. It was unlikely to be exceeded. As the use of these early microprocessors became more widespread, applications started to grow in size and the use of operating systems like CP/M and high level languages increased memory requirements until the address range started to limit applications. Various techniques like bank switching and program overlays were developed to help.

Bank switching simply involves having several banks of memory with the same address locations. At any one time, only one bank of memory is enabled and accessible by the microprocessor. Bank selection is made by driving the required bank select line. These lines come from either an external parallel port or latch whose register(s) appear as a bank switching control register within the processors's normal address map. In the example, the binary value 1000 has been loaded into the bank selection register. This asserts the select line for bank 'a' which is subsequently accessed by the processor. Special VLSI (very large scale integration) parts were even developed which provided large number of banks and additional control facilities: the Motorola MC6883 SAM is a well-known example used in the Dragon MC6809–based home computer from the early 1980s.

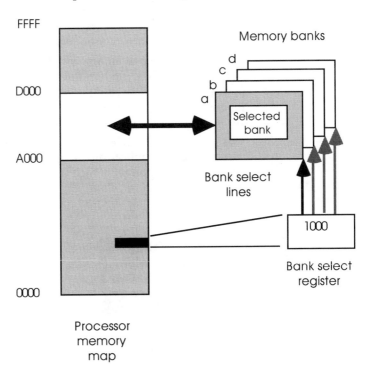

Bank switching

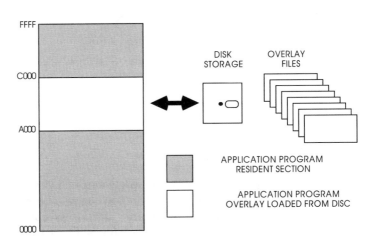

Program overlays

Program overlays are similar to bank switching except that some form of mass storage is used to contain the different overlays. If a particular subroutine is not available, the software stores part of its memory as a file on disk and loads a new program section from disk into its place. Several hundred kilobytes of program, divided into smaller sections, can be made to overlay a single block of memory, making a large program fit into the 64 kilobyte memory.

System integrity

Another disadvantage with this type of architecture is their unpredictability in handling error conditions. A bug in a software application could corrupt the whole system, causing a system to either crash, hang up or, even worse, perform some unforeseen operations. The reasons are quite simple: there is no partitioning between data and programs within the architecture. An application can update a data structure using a corrupt index pointer which overwrites a part of its program.

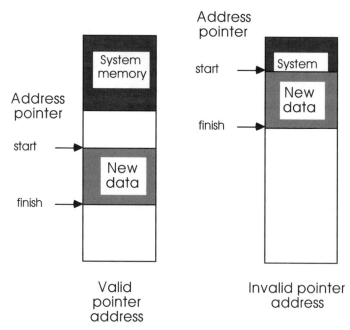

*System corruption via
an invalid pointer*

Data are simply bytes of information which can be interpreted as instruction codes. The processor calls a subroutine within this area, starts to execute the data as code and suddenly the whole system starts performing erratically! On some machines, certain undocumented code sequences could put the processor in a test mode and start cycling through the address ranges etc. These attributes restricted their use to non-critical applications.

It is important to put these characteristics into perspective: at the time (late 1970s) processing power offered at the low cost was revolutionary and was more than sufficient for many applications. Devices were at their integration limits using the then current state of the art technology. While the deficiencies were quite apparent, better designs could only be realised using discrete implementations and mini-computers at a lot higher cost. It was the combination of reasonable performance and at a very low cost that ignited the revolution. Many applications were waiting for that balance to be achieved.

In terms of numbers, 8 bit processors like the Zilog Z80, the Intel 8080 and the Motorola MC6800 and their controller derivatives have sold more units than any of the more powerful microprocessors. Millions are used to control car ignition systems, television remote control units, domestic appliances etc. The list is endless. In terms of the performance race, these architectures have been left far behind — but these machines were the acorns from which today´s super micros have grown.

Requirements for a new processor architecture

The deficiencies of the the the 8 bit processors set the goals for the new generation of processors that would use the higher integration capabilities rapidly being developed. The central aim was to provide more and more powerful facilities.

The new architecture would need:

* A larger data size of at least 16 bits.

  An obvious statement, yet is 16 bits really enough? For many computational applications, 32 bits would be a better choice. Does the external bus size need to be the same as the internal size? If a 32 bit external bus is used, is there a package with enough pins?

* More registers with more general purpose use.

  Another obvious requirement. The questions are how many, how big and how general purpose? Should the registers form a superset of an existing 8 bit architecture, and thus provide some form of compatibility?

* A larger memory addressing range.

  Great, but how big? 2, 4, 8 or 16 times larger? Should it be a group of 64 Kbyte blocks? How long will it be before that is insufficient?

- More complex and powerful instructions and addressing modes.

  Another winner! With more powerful instructions, the
  less work compiler writers need to do the more efficient
  compilers will be. The more work an instruction can
  do, the less that are needed and, therefore, the required
  system memory and bus bandwidth will be smaller.
  However, the instruction size needs to increase to
  provide sufficient unique codes.

  At this time, there were even suggestions that the
  native processor instruction set should be a set of high
  level language primitives with no assembler level support. The counter-argument stated that many of the
  simple operations performed by such primitives could
  be split into several simpler lower level instructions
  which would execute faster on machines with such
  lower level support. This argument reappeared as part
  of the RISC processor philosophy.

- System partitioning and protection.

  The processor would require resource protection mechanisms to prevent system crashes and allow the orderly
  recovery or shutting down of the system. Many control
  applications could not tolerate previous processor ec-
  centricities.

It was against this background of larger data sizes,
more powerful and complex instructions, ever increasing
hardware support for compiler writers and the need for more
efficient memory utilisation and sophistication, that Motorola
designed its MC68000 CISC processor architecture.

Software compatibility

When faced with the challenge of designing a new
microprocessor architecture, one of the most critical dilem-
mas is how compatible must it be with preceding designs or
families. Does compatibility constrain other developments
and ideas which will prevent further developments? Would
the best solution be a clean sheet coupled with a crystal ball to
identify features that will be needed in ten years time? This
was the problem faced by Motorola and the team behind the
MACSS (Motorola´s advanced computer system on silicon)
project.

The company continued development of its microprocessors and introduced the MC6809 8 bit processor in the late
1970s. This design took the basic MC6800 architecture and

extended it. These extensions provided answers to many of the criticisms. The programmers register model shows the changes:

- The old A and B accumulators have been joined to create a new D register, while still preserving compatibility with the MC6800 model. This register provided the much needed 16 bit support.
- There are two new 16 bit registers: an additional index register and a user stack pointer.
- An 8 bit direct page register appeared to allow addressing into any 256 byte direct page within the 64 Kbyte memory map. This address register selects the page location within it.

The instruction set and addressing modes were based on the MC6800 with new additions:

- 16 bit compare and addition instructions were added.
- Addressing modes were expanded to include larger offsets and indirection. Indirection allowed the contents of a memory address to be used for an address rather than the address itself.

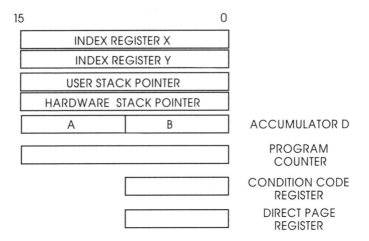

*The MC6809
programmer's model*

The dilemma was quite simple: should the MC6809/6800 model be extended further or start afresh? There were already hints that software compatibility would not be an unquestioned right: software was beginning to be developed in high level languages such as PASCAL, which removed the need to code in assembler and provided a simple migration path from one architecture to another via recompilation. The question still remained unanswered: extend or start from scratch?

Enter the MC68000

The route that Motorola took with the MC68000 can be summed up with the advertising slogan used at its launch in August 1979:

'Break away from the Past'

The MC68000 was a complete design from scratch with the emphasis on providing an architecture that looked forward without the restrictions of remaining compatible with past designs. The only support for the old MC6800 family was a hardware interface to allow the new processor to use the existing M6800 peripherals while new M68000 parts were being designed.

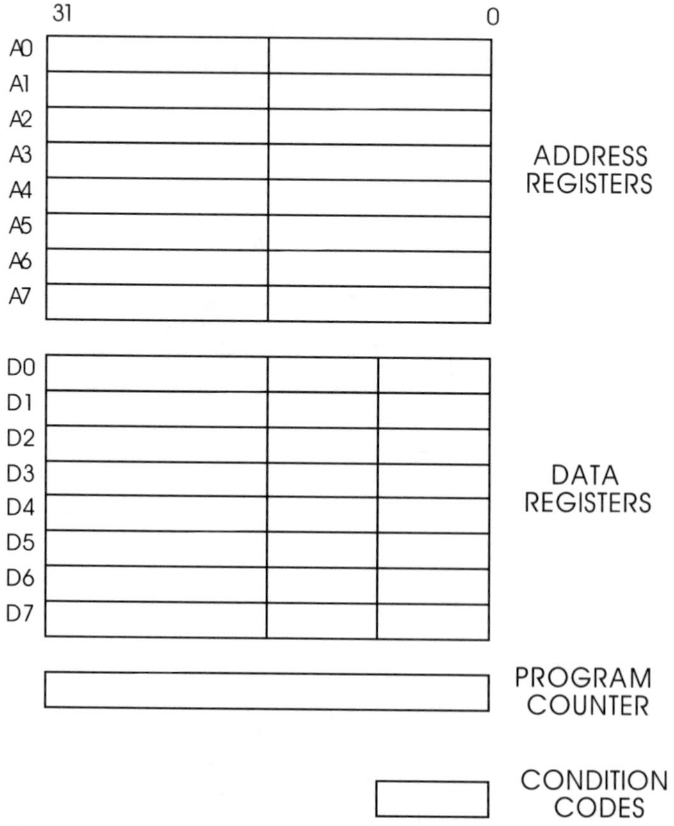

```
      31                                    0
AO  ┌──────────────┬──────────────┐
A1  ├──────────────┼──────────────┤
A2  ├──────────────┼──────────────┤
A3  ├──────────────┼──────────────┤      ADDRESS
A4  ├──────────────┼──────────────┤      REGISTERS
A5  ├──────────────┼──────────────┤
A6  ├──────────────┼──────────────┤
A7  └──────────────┴──────────────┘

DO  ┌──────────┬──────────┬───────┐
D1  ├──────────┼──────────┼───────┤
D2  ├──────────┼──────────┼───────┤
D3  ├──────────┼──────────┼───────┤      DATA
D4  ├──────────┼──────────┼───────┤      REGISTERS
D5  ├──────────┼──────────┼───────┤
D6  ├──────────┼──────────┼───────┤
D7  └──────────┴──────────┴───────┘

    ┌──────────────────────────────┐     PROGRAM
    └──────────────────────────────┘     COUNTER

                      ┌───────────┐       CONDITION
                      └───────────┘       CODES
```

The MC68000 USER
programmer's model

Its design took many of the then current mini and mainframe computer architectural concepts and developed them using VLSI silicon technology. The programmer's register model shows how dramatic the change was. Gone are the dedicated 8 and 16 bit registers to be replaced by two groups of eight data registers and eight address registers. All these registers and the program counter are 32 bits wide. The processor contained about 68000 transistors — about 10 times the number used by its predecessors. Folklore, if there is such a thing within a 30 year old industry, states that the number of transistors gave the new processor its part number, and not that 68000 was a natural successor to the MC6800 numbering. It is suspected that the transistor count was coincidental, but marketing is never slow to realise an opportunity!

Complex instructions, microcode and nanocode

The MC68000 microprocessor has a complex instruction set, where each instruction is very powerful in its processing capability and supports large numbers of registers and addressing modes. The instruction format consists of an op code followed by a source effective address and a destination effective address. To provide sufficient coding bits, the op code is 16 bits in size with further 16 bit operand extensions for offsets and absolute addresses. Internally, the instruction does not operate directly on the internal resources, but is decoded to a sequence of microcode instructions, which in turn calls a sequence of nanocode commands which controls the sequencers and arithmetic logic units (ALU). This is analogous to the many macro subroutines used by assembler programmers to provide higher level 'pseudo' instructions. On the MC68000, microcoding and nanocoding allow instructions to share common lower level routines, thus reducing the hardware needed and allowing full testing and emulation prior to fabrication. Neither the microcode nor the nanocode sequences are available to the programmer.

These sequences, together with the sophisticated address calculations necessary for some modes, often take more clock cycles than are consumed in fetching instructions and their associated operands from external memory. This multilevel decoding automatically lends itself to a pipelined approach which also allows a prefetch mechanism to be employed.

Pipelining works by splitting the instruction fetch, decode and execution into independent stages: as an instruction goes through each stage, the next instruction follows it without waiting for it to completely finish. If the instruction fetch is included within the pipeline, the next instruction can be read from memory, while the preceding instruction is still being executed as shown.

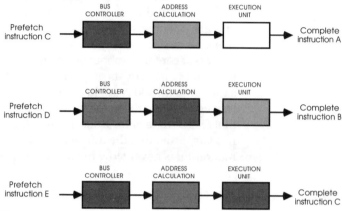

Pipelining instructions

The only disadvantage with pipelining concerns pipeline stalls. These are caused when any stage within the pipeline cannot complete its allotted task at the same time as its peers as shown. This can occur when wait states are inserted into external memory accesses, instructions use iterative techniques or there is a change in program flow.

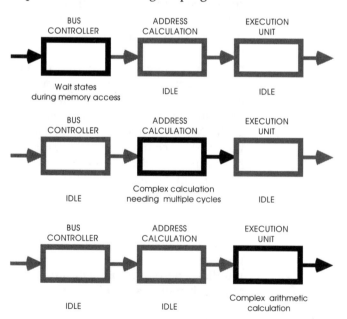

Pipeline stalls

With iterative delays, commonly used in multiply and divide instructions and complex address calculations, the only possible solutions are to provide additional hardware support, add more stages to the pipeline, or simply suffer the delays on the grounds that the performance is still better than anything else! Additional hardware support may or may not be within a designer´s real estate budget (real estate refers to the silicon die area, and directly the number of transistors available). Adding stages also consumes real estate and increases pipeline stall delays when branching.

The main culprits are program branching and similar operations. The problem is caused by the decision whether to take the branch or not being reached late in the pipeline, i.e. after the next instruction has been prefetched. If the branch is not taken, this instruction is valid and execution can carry on. If the branch is taken, the instruction is not valid and the whole pipeline must be flushed and reloaded. This causes additional memory cycles before the processor can continue. The delay is dependent on the number of stages, hence the potential difficulty in increasing the number of stages to reduce iterative delays. This interrelation of engineering trade-offs is a common theme within microprocessor architectures. Similar problems can occur for any change of flow: they are not limited to just branch instructions and can occur with interrupts, jumps, software interrupts, etc. With the large usage of these types of instructions, it is essential to minimise these delays. The longer the pipeline, the greater the potential delay.

The MC68000 hardware

The device first appeared running at 4 MHz in a 64 pin dual-in-line package. Fabricated in NMOS (N channel metal oxide semiconductor), its 68,000 transistors were etched on to a die 0.246 inches by 0.261 inches.

The processors signals can be divided into several groups.

Address bus

The address bus, signals A1 to A23, is non-multiplexed and 24 bits wide, giving a single linear addressing space of 16 Mbytes. A0 is not brought out directly but is internally decoded to generate upper and lower data strobes. This allows the processor to access either or both the upper and lower bytes that comprise the 16 bit data bus.

Data bus

The data bus, D0 to D15, is also non-multiplexed and provides a 16 bit wide data path to external memory and peripherals. The processor can use data in either byte, word

(16 bit) or long word (32 bit) values. Both word and long word data is stored on the appropriate boundary, while bytes can be stored anywhere. The diagram shows how these data quantities are stored. All addresses specify the byte at the start of the quantity. If an instruction needs 32 bits of data to be accessed in external memory, this is performed as two successive 16 bit accesses automatically.

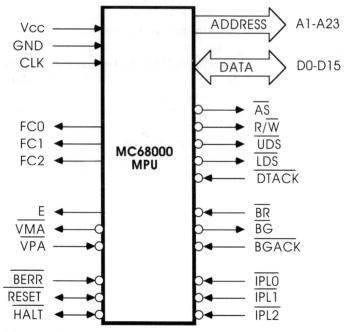

The MC68000 pinout

Instructions and operands are always 16 bits in size and accessed on word boundaries. Attempts to access instructions, operands, words or long words on odd byte boundaries cause an internal 'address' error and exception.

Function codes

The function codes, FC0-FC2, provide extra information describing what type of bus cycle is occurring. These codes and their meanings are shown in the table. They appear at the same time as the address bus data and indicate program/data and supervisor/user accesses. In addition, when all three signals are asserted, the present cycle is an interrupt acknowledgement, where an interrupt vector is passed to the processor. Many designers use these codes to provide hardware partitioning. This technique is described later in Chapter 9.

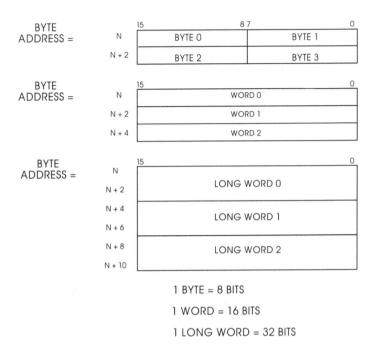

1 BYTE = 8 BITS

1 WORD = 16 BITS

1 LONG WORD = 32 BITS

MC68000 data organization

Function code			Reference class
FC0	FC1	FC2	
0	0	0	Reserved
0	0	1	User data
0	1	0	User program
0	1	1	Reserved(I/O space)
1	0	0	Reserved
1	0	1	Supervisor data
1	1	0	Supervisor program
1	1	1	CPU space/ interrupt ack

The MC68000 function codes and their meanings

MC68000 asynchronous bus

The MC68000 bus is fundamentally different to the buses used on the MC6800 and MC6809 processors. Their buses were synchronous in nature and assumed that both memory and peripherals could respond within a cycle of the bus. The biggest drawback with this arrangement concerned system upgrading and compatibility. If one component was uprated, the rest of the system needed uprating as well. It was for this reason that all M6800 parts had a system rating built into their part number. If a design specified a MC6809B, then it needed 2 MHz parts and subsequently, could not use an 'A' version which ran at 1 MHz. If a design based around the 1 MHz processor and peripherals was upgraded to 2 MHz, all the parts would need replacing. If a peripheral was not available at the higher speed, the system could not be upgraded. With the increasing processor and memory speeds, this restriction was unacceptable.

The MC68000 bus is truly asynchronous: it reads and writes data in response to inputs from memory or peripherals which may appear at any stage within the bus cycle. Providing certain signals meet certain setup times and minimum pulse widths, the processor can talk to anything. As the bus is truly asynchronous it will wait indefinitely if no reply is received. This can cause similar symptoms to a hung processor, however, most system designs use a watchdog timer and the processor bus error signal to resolve this problem.

A typical bus cycle starts with the address, function codes and the read/write line appearing on the bus. Officially, this data is not valid until the address strobe signal AS* appears but many designs start decoding prior to its appearance and use the AS* to validate the output. The upper and lower data strobes, together with the address strobe signal (both shown as DS*) are asserted to indicate which bytes are being accessed on the bus. If the upper strobe is asserted, the upper byte is selected. If the lower strobe is asserted, the lower byte is chosen. If both are asserted together, a word is being accessed.

Once complete, the processor waits until a response appears from the memory or peripheral being accessed. If the rest of the system can respond without wait states (i.e. the decoding and access times will be ready on time) a DTACK* (Data Transfer ACKnowledge) signal is returned. This occurs slightly before clock edge S4. The data is driven on to the bus, latched and the address and data strobes removed to acknowledge the receipt of the DTACK* signal by the processor. The system responds by removing DTACK* and the cycle is complete. If the DTACK* signal is delayed for any reason, the processor will simply insert wait states into the cycle. This allows extra time for slow memory or peripherals to prepare data.

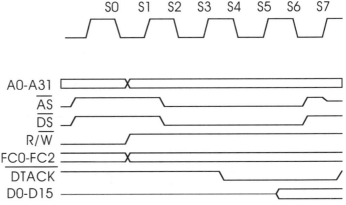

A MC68000 asynchro-nous bus cycle

The advantages that this offers are manyfold. First, the processor can use a mixture of peripherals with different access speeds without any particular concern. A simple logic circuit can generate DTACK* signals with the appropriate delays as shown. If any part of the system is upgraded, it is a simple matter to adjust the DTACK* generation accordingly. Many M68000 boards provide jumper fields for this purpose and a single board and design can support processors running at 8, 10, 12 or 16 MHz. Secondly, this type of interface is very easy to interface to other buses and peripherals. Additional time can be provided to allow signal translation and conversion.

M6800 synchronous bus

Support for the M6800 synchronous bus initially offered early M68000 system designers access to the M6800 peripherals and allowed them to build designs as soon as the processor was available. With today's range of peripherals with specific M68000 interfaces, this interface is less used. However, the M6800 parts are now extremely inexpensive and are often used in cost-sensitive applications.

The additional signals involved are the E clock, valid memory address VMA* and valid peripheral address VPA*. The cycle starts in a similar way to the M68000 asynchronous interface except that DTACK* is not returned. The address decoding generates a peripheral chip select which asserts VPA*. This tells the M68000 that a synchronous cycle is being performed.

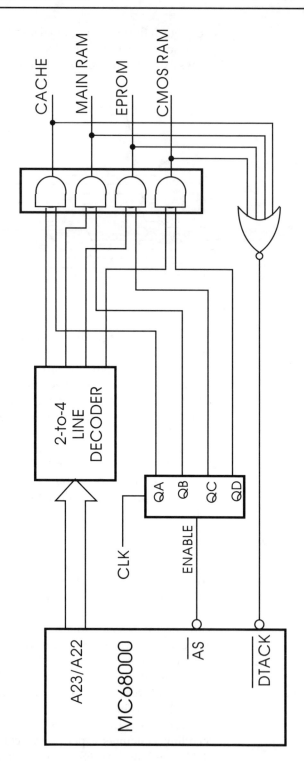

Example DTACK*
generation

The address decoding monitors the E clock signal, which is derived from the main system clock, but is divided down by 10 with a 6:4 mark/space ratio. It is not referenced from any other signal and is free running. At the appropriate time (i.e. when E goes low) VMA* is asserted. The peripheral waits for E to go high and transfers the data. When E goes low, the processor negates VMA* and the address and data strobes to end the cycle.

For systems running at 10 MHz or lower, standard 1 MHz M6800 parts can be used. For higher speeds, 1.5 or 2 MHz versions must be employed. However, higher speed parts running at a lower clock frequency will not perform the peripheral functions at full performance.

Interrupts

Seven interrupt levels are supported and are encoded on to three interrupt pins IP0-IP2. With all three signals high, no external interrupt is requested. With all three asserted, a non-maskable level 7 interrupt is generated. Levels 1–6, generated by other combinations, can be internally masked by writing to the appropriate bits within the status register.

The interrupt cycle is started by a peripheral generating an interrupt. This is usually encoded using a 148 priority encoder. The appropriate code sequence is generated and drives the interrupt pins. The processor samples the levels and requires the levels to remain constant to be recognised. It is recommended that the interrupt level remains asserted until its interrupt acknowledgement cycle commences to ensure recognition. Once the processor has recognised the interrupt, it waits until the current instruction has been completed and starts an interrupt acknowledgement cycle. This starts an external bus cycle with all three function codes driven high to indicate an interrupt acknowledgement cycle.

The interrupt level being acknowledged is placed on address bus bits A1–A3 to allow external circuitry to identify which level is being acknowledged. This is essential when one or more interrupt requests are pending. The system now has a choice over which way it will respond :

• If the peripheral can generate an 8 bit vector number, this is placed on the lower byte of the address bus and DTACK* asserted. The vector number is read and the cycle completed. This vector number then selects the address and subsequent software handler from the vector table.

• If the peripheral cannot generate a vector, it can assert VPA* and the processor will terminate the cycle using the M6800 interface. It will select the

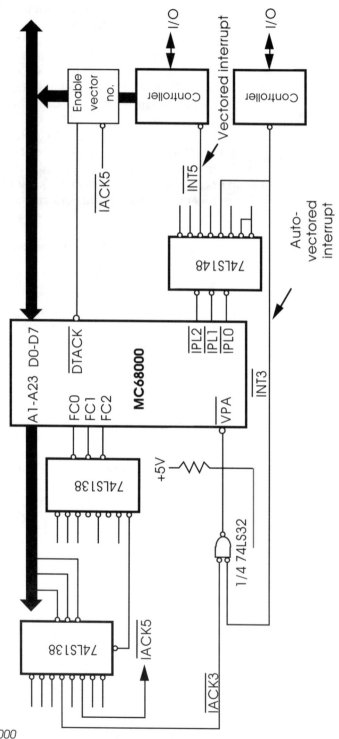

*An example MC68000
interrupt design*

specific interrupt vector allocated to the specific interrupt level. This method is called auto-vectoring.

To prevent an interrupt request generating multiple acknowledgements, the internal interrupt mask is raised to the interrupt level, effectively masking any further requests. Only if a higher level interrupt occurs will the processor nest its interrupt service routines. The interrupt service routine must clear the interrupt source and thus remove the request before returning to normal execution. If another interrupt is pending from a different source, it will be recognised and cause another acknowledgement to occur.

A typical circuit is shown. Here, level 5 has been allocated as a vectored interrupt and level 3 autovectored. The VPA* signal is gated with the level 3 interrupt to allow level 3 to be used with vectored or autovectored sources in future designs.

Error recovery and control signals

There are three signals associated with error control and recovery. The bus error BERR*, HALT* and RESET* signals can provide information or be used as inputs to start recovery procedures in case of system problems.

The BERR* signal is the counterpart of DTACK*. It is used during a bus cycle to indicate an error condition that may arise through parity errors or accessing non-existent memory. If BERR* is asserted on its own, the processor halts normal processing and goes to a special bus error software handler. If HALT* is asserted at the same time, it is possible to rerun the bus cycle. BERR* is removed followed by HALT* one clock later, after which the previous cycle is rerun automatically. This is useful to screen out transient errors. Many designs use external hardware to force a rerun automatically but will cause a full bus error if an error occurs during the rerun.

Without such a signal, the only recourse is to complete the transfer, generate an immediate non-maskable interrupt and let a software handler attempt to sort out the mess! Often the only way out is to reset the system or shut it down. This makes the system extremely intolerant of signal noise and other such transient errors.

The RESET* and HALT* signals are driven low at power up to force the MC68000 into its power up sequence. The operation take about 100 ms, after which the signals are negated and the processor accesses the Reset vector at location 0 in memory to fetch its stack pointer and program counter from the two long words stored there.

RESET* can be driven by the processor in response to the RESET instruction. This allows the processor to reset external hardware without resetting itself.

The HALT* signal can be driven by the processor in response to a catastrophic failure or externally to halt the processor. A typical catastrophic cause may be a bus error encountered during stack frame building.

Bus arbitration

The MC68000 has a three-wire bus arbitration scheme. These signals are used in the bus arbitration process to allow another bus master, such as a DMA controller or another processor, to request and obtain the bus. While the bus is arbited away, the MC68000 bus signals are tristated. This mechanism is a more elegant replacement for the clock stretching and cycle stealing techniques used with many 8 bit processors which relied on delaying clock edges from the processor so that it did not start another cycle.

The arbitration scheme allocates the MC68000 a lower bus priority than an external request. An external master must request the bus prior to it starting a bus cycle by asserting the bus request signal BR*. The processor completes its current bus cycle, tristates its bus and control signals and asserts the bus grant signal BG*. This informs the other master that the bus is available as far as the processor is concerned. The master must check that the current cycle is complete (AS*, DTACK*, etc. negated) before asserting bus grant acknowledge BGACK* and negating BR*. The bus is now available for use.

If BR* is negated prior to BGACK* being asserted, the processor assumes that the request is erroneous and takes back the bus. Apart from the obvious applications with multi-processing systems and DMA control, this method can be used to allow in-circuit emulators to be used in systems without removing the processor.

Typical system

The diagram shows a typical MC68000 system. There is no differentiation between I/O and memory accesses: all peripherals, such as serial and parallel ports, appear as locations within the memory map and the same instructions are used to access both memory and I/O.

The memory is constructed from two byte wide memory banks forming a 16 bit data path. The MC68901 MultiFunction Peripheral has an 8 bit data bus and is connected to either the odd or even byte. Its registers therefore appear on consecutive odd or even addresses and not simply successive addresses.

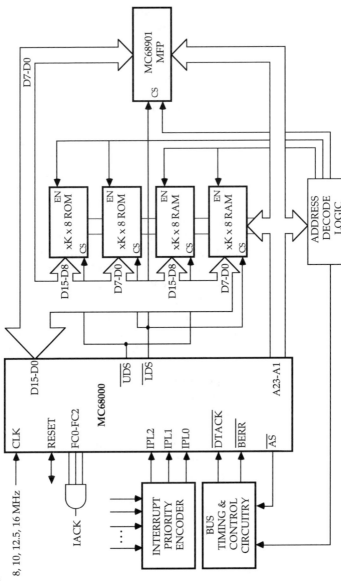

A typical MC68000 system

Just as the external hardware interface was quite different from anything previously seen, the internal architecture was a complete departure from previous designs: the architecture had more in common with a Digital Equipment Corporation PDP-11 mini-computer than a microprocessor! The first fundamental difference was the register set.

The register set

The MC68000 register set supports 8, 16 and 32 bit wide data as standard and, with the large number of registers, allows data to be stored internally rather than on an external stack, with the benefit of reduced stack manipulation. This was important, considering that internal register storage was faster to access and thus offered quicker instruction execution. With a complex iterative process, the ability to store data internally, so that after initial loading any further data fetches were removed from the execution time, was an obvious benefit. If the system had had to suffer wait states when accessing memory, this benefit increased.

The general purpose nature of the data and address registers further improves performance: whatever operation can be performed on data register 1 can be performed on any of the other seven. Similarly, any addressing mode can be used with any one of the eight address registers. All the overhead of moving data to and from a specific register for address calculations or manipulation is removed. This overhead can be as high as 40% on the total number of executed instructions within a dedicated register approach, although it is application dependent.

The condition code register looks very similar to that found within an 8 bit processor, except that almost every instruction execution updates it automatically without having to perform a separate compare operation. This is beneficial in performing branch determination and allows single instructions to test for a condition and decide to change the program flow.

The program counter is specified as a full 32 bits, but only the lower 24 bits are bonded out to provide a 24 bit external address bus. In addition to the user registers there are some additional supervisor registers.

31	16	15	0	
INTERRUPT STACK POINTER				A7'
MC68000		STATUS REGISTER		SR

The supervisor registers for the MC68000

The USER/SUPERVISOR concept

Many earlier 8 bit microprocessor systems could easily be crashed by bad programming and/or software errors, usually associated with application tasks rather than the operating systems that were being used. The reason was quite simple: there was little or no separation or protection of the controlling software and application tasks running under it.

To solve this deficiency, the MC68000 family uses the concept of two separate states with separate and protected resources. The processor can either be halted or executing in the USER or SUPERVISOR state. It normally executes code in the USER state but when an exception occurs, it switches into a SUPERVISOR state with its own separate stack pointer, A7′, and access to the processor control registers. The SUPERVISOR stack pointer is used for the stack frames that are generated as part of the exception process. They contain essential information to restore the program flow prior to the exception and allow nesting, given that there is sufficient stack memory allocated to store all the generated frames.

The transition can be caused by a processing error, external interrupts or a software interrupt TRAP instruction. The duplicated stack pointer allows separate stacks to be maintained and also allows application software, running in USER mode, to cause errors without destroying an operating system kernel running in SUPERVISOR mode. Using these two states permits application and operating system software to be separated, preventing rogue applications from corrupting the whole system.

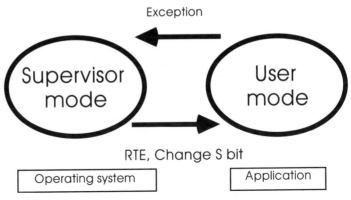

The M68000 USER/
SUPERVISOR concept

However, the processor′s dual status needs to be repeated in the external design. If the whole of the external memory was available to both the USER and SUPERVISOR states, each program would be put at risk from corruption. The memory system must be partitioned so that programs can be protected. This function is performed by the three function code signals referenced in the previous hardware interface description. The three pins offer eight codes which indicate the type of transfer that is taking place, i.e. program or data, SUPERVISOR or USER, or interrupt acknowledge/CPU space cycle. By taking these outputs and including them in the address decode logic, blocks of memory can be reserved for specific use and can therefore be protected.

Once in SUPERVISOR mode, the exception is serviced via a specific exception handler, depending on what caused the exception to occur. Up to 255 x 32 bit pointers are stored in a vector table stored in the bottom 1,024 bytes of memory.

Exceptions and the vector table

As described previously, the MC68000 processor responds to external interrupts and bus errors by treating them as an exception, switching into supervisor mode, selecting the appropriate vector and executing its software exception handler. Due to the internal microcoding within it, the MC68000 can identify further error and potential error conditions caused by incorrect software or hardware failures and treat them as separate entities rather than a general fault condition. This allows fault conditions to be identified early, and gives a far better indication as to the problem. The following diagram shows the vector table and the conditions associated with each vector.

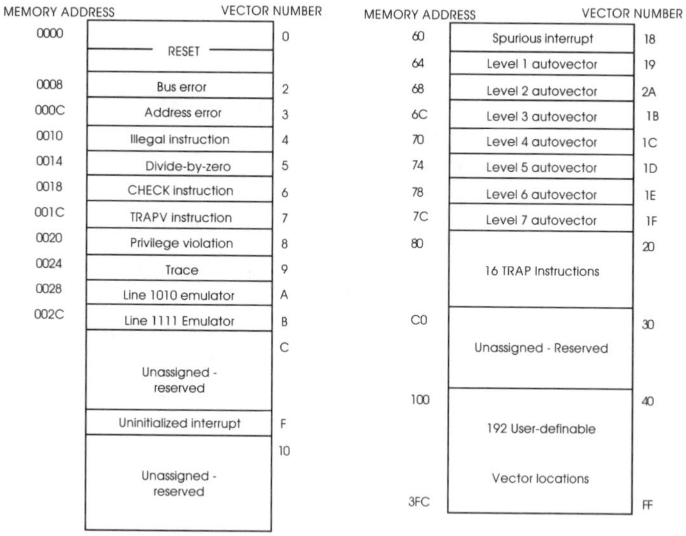

MEMORY ADDRESS		VECTOR NUMBER
0000	RESET	0
0008	Bus error	2
000C	Address error	3
0010	Illegal instruction	4
0014	Divide-by-zero	5
0018	CHECK instruction	6
001C	TRAPV instruction	7
0020	Privilege violation	8
0024	Trace	9
0028	Line 1010 emulator	A
002C	Line 1111 Emulator	B
		C
	Unassigned - reserved	
	Uninitialized interrupt	F
		10
	Unassigned - reserved	

MEMORY ADDRESS		VECTOR NUMBER
60	Spurious interrupt	18
64	Level 1 autovector	19
68	Level 2 autovector	2A
6C	Level 3 autovector	1B
70	Level 4 autovector	1C
74	Level 5 autovector	1D
78	Level 6 autovector	1E
7C	Level 7 autovector	1F
80		20
	16 TRAP Instructions	
C0		30
	Unassigned - Reserved	
100		40
	192 User-definable	
	Vector locations	
3FC		FF

The MC68000 exception table

The first entry in the table contains the RESET vector used by the processor on power up. The bus error vector is followed by the address error. This is selected if an attempt is made to fetch an instruction or word on an odd byte boundary. The next location is used for any attempt to execute an illegal instruction. The processor detects any such attempts and will not go into an undocumented state. This vector is often used to implement breakpoints: the original instruction is overwritten by an illegal one, which on execution, passes control to the exception handler. Divide by zero is self explanatory. The next two vectors are accessed if certain conditions are met. The TRAPV instruction causes this handler to be invoked if the V flag within the condition code register is set, i.e. when an overflow condition exists. The CHECK handler is used if data is found to be out of bounds as a result of a CHECK instruction.

The privilege violation handler is invoked if the processor in USER mode attempts to perform a SUPERVISOR operation, such as changing the interrupt mask. Within the handler, the SUPERVISOR can decide whether to perform the operation or not. The TRACE exception vector provides processor debug facilities: if the T bit is set in the SUPERVISOR status register, the processor takes the TRACE exception after each instruction is completed. This allows a debugger to single step through a program.

The next pair of vectors are used when certain types of instructions are executed. These are the famous F-line and A-line instructions. Their names come from the bit pattern encoded into the four highest-order bits within the instruction. None of the standard instructions use that encoding, thus allowing programmers to define their own additional A-line and F-line instructions, which will invoke their own exception handler during execution. The handler can simulate the instruction execution. Motorola used the F-line mechanism for its floating point coprocessor instructions, so that software could perform the calculations if no hardware support was available. The A-line vector is often used for breakpoints as an alternative to the illegal instruction vector.

Vector F, sandwiched between two banks of reserved vectors, handles errors during the hardware interrupt sequence when an incorrect vector is passed to the processor. The spurious interrupt vector which follows it is similar: it is selected when a bus error signal terminates the interrupt acknowledge cycle. The next seven locations are for the autovectored interrupts, with each interrupt level assigned its own handler.

The 16 TRAP vectors are next. They are the equivalent of a software interrupt and allow USER code to make requests of the SUPERVISOR. Operating system calls are made in this way and commands like TRAP #1 and TRAP #2 are very common. Parameters are often passed via a register to further define the request.

After another bank of reserved vectors, there are 192 allocated for the user to define. This may seem to be an overkill, but it does allow a group of different handlers to be assigned to a particular peripheral, providing a faster response. This is advantageous with peripherals that generate vector numbers. With a multiprotocol serial port, one vector could be assigned to handle interrupts when the port is in asynchronous mode, a second when in SDLC, a third for bisync support and so on. When the port is configured, the vector register is loaded to redirect the interrupt to the correct handler. This removes the overhead of determining what mode the device is in and subsequently how to handle the interrupt. Similar techniques can be used to dynamically reallocate interrupt handlers for particular ports. An operating system may decide to take a printer port offline by simply reprogramming its vector register, thus changing the vector number it generates to request servicing. Instead of being handled by a device driver, an error handler can take over and simply discard any data, thus taking it offline.

It is important to remember that any vector number from 0 to 255 can be passed during an interrupt acknowledgement cycle. It is prudent to have all vectors assigned to a handler. If unused vectors are set to non-existent or random memory locations, any external hardware faults or software errors could cause the processor to halt without providing any information as to the cause. Once the MC68000 has halted, only a system reset can start it again with the subsequent data loss.

Addressing modes

The MC68000 has a far greater range of addressing modes than its predecessors as shown in the list of their forms and names. The reason for this expansion is to reduce the number of calculations needed to locate and process data. The (An)+ and -(An) modes perform automatic post-increment and pre-decrement of their respective register contents. This allows pointers to be used to access data stored in a table and not have to perform register increment or decrement instructions after every access. Another advantage of these modes is their automatic size adjustment. If the pointer is used to access a long word, it is incremented by 4 to position it at the next valid byte location. Similarly, word and byte accesses are adjusted by 2 and 1, respectively. These modes reduce the number of instructions that an 8 bit architecture without these facilities would need.

The program counter relative addressing modes allow the easy development of relocatable and position independent code. If an instruction specifies one of these modes, an offset from the data location relative to the current program counter is encoded into an operand. If the program and the data areas are then moved into a different memory location, while retaining the same offset, the data can still be located correctly. If complete programs are written in such manner, multitasking can be achieved without hardware memory management.

Dn	Data register direct
An	Address register direct
(An)+	Address reg. indirect w/ post-increment
-(An)	Address reg. indirect w/ pre-decrement
d(An)	Displaced address register indirect
d(An,Rx)	Indexed, displaced address reg. indirect
d(PC)	Program counter relative
d(PC,Rx)	Indexed program counter relative
#xxxxxxxx	Immediate
$xxxx	Absolute short
$xxxxxxxx	Absolute long

The MC68000 address-ing modes

Instruction set

Each instruction mnemonic has a single letter suffix which defines the data size:

- .B for byte,
- .W for a 16 bit word, and
- .L for a 32 bit long word.

Most assemblers assume word data size if no suffix is specified. Each instruction is based around a 16 bit format with optional operands depending on selected addressing modes. It takes a diadic form using two effective addresses to form two sources AND destination, i.e. if two registers are multiplied together, the result is stored in one of them, overwriting the original value. However, it must be noted that although the encoding appears regular (i.e. bits 9 to 11 specify a source register and bits 0 to 5 specify a destination effective address) there are many instructions where this regularity is not observed such as the CMPI (compare immediate), LSL and LSR logical shift instructions. This greatly increases the decoding complexity and, as will be seen later on, became a major criticism of CISC architectures like the M68000 family.

ABCD	Bcc	DIVS	MOVE	OR	SBCD
ADD	BCHG	DIVU	MOVE to	ORI	Scc
ADDA	BCLRB	EOR	CCR	ORI to	STOP
ADDI	RA	EORI	MOVE SR	CCR	SUB
ADDQ	BSET	EORI to	MOVE	ORI to	SUBA
ADDX	BSR	CCR	USP	SR	SUBI
AND	BTSTC	EORI to	MOVEA	PEA	SUBQ
ANDI	HK	SR	MOVEM	RESET	SUBX
ANDI to CCR	CLR	EXG	MOVEP	ROL	SWAP
ANDI to SR	CMP	EXT	MOVEQ	ROR	TAS
ASL	CMPA	JMP	MULU	ROXL	TRAP
ASR	CMPIC	JSR	MULS	ROXR	TRAPV
	MPMD	LEA	NEG	RTE	TST
	Bcc	LINK	NEGX	RTR	UNLK
		LSL	NOP	RTS	
		LSR	NOT		

*The MC68000
instruction set*

With all these new and powerful commands, addressing modes and sophisticated hardware, CISC processors like the M68000 family catalysed the migration of software techniques and applications from mini and mainframe computers to the microprocessor arena.

Multitasking operating systems

Perhaps the most apparent reaction to this high performance CISC architecture was the sudden appearance of sophisticated multitasking operating systems for microprocessors. Most 8 bit architectures could offer operating system support, such as CP/M for 8080 and Z80 families and MDOS for the MC6800 and MC6809 processors, but these were restricted in memory size and system response speeds and could only execute a single task at a time. Code development was often performed in assembler or BASIC with very rudimentary debugging. There was no system protection, with application errors often crashing the system and thus further hindering debugging investigations.

The MC68000 offered several solutions. Its linear 16 Mbyte memory addressing space obviously removed any memory limitations, providing sufficient memory could be afforded or installed. Its linear addressing, coupled with the addressing mode support for relocatable code, allowed multiple programs or tasks to be relocated into memory, even though they originally may have been designed to run at the same nominal location. Without this support, such relocation could only be performed with a memory management unit which, at this time, inserted delays within the bus cycles and increased software overhead. The Intel 80x86 architecture took a different approach and used a segmentation scheme based on 64 Kbyte segments. This allowed them to preserve compatibility with their previous 8 bit architectures, although adding software overhead to recognise 64 Kbyte boundaries and reprogram segment registers. This was a trade-off that Motorola did not agree with.

With the ability to load and hold multiple tasks, all that was required to make a multitasking operating system was a timer to control a scheduler or kernel.

A multitasking operating system works by dividing the processors' time into discrete time slots.

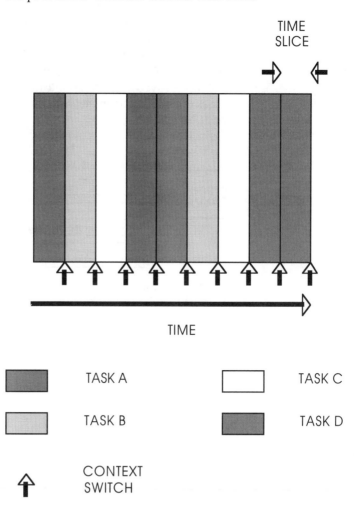

Time slicing

Each application or task requires a certain number of time slots to complete its execution. The operating system kernel decides which task can have the next slot, so instead of a task executing continuously until completion, its execution is interleaved with other tasks. This sharing of processor time between tasks gives the illusion to each user that he is the only one using the system.

Context switching, task tables and kernels

Multitasking operating systems are based around a multitasking kernel which controls the time slicing mechanisms. A time slice is the time period each task has for execution before it is stopped and replaced during a context switch. This is periodically triggered by a hardware interrupt from the system timer. This interrupt may provide the system clock, and several interrupts may be executed and counted before a context switch is performed. When a context switch is performed, the current task is interrupted, the processor's registers are saved in a special table for that particular task and the task is placed back on the 'ready' list to await another time slice.

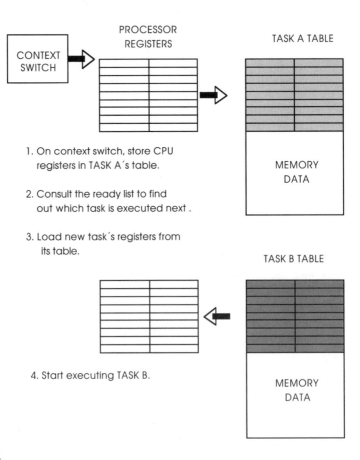

1. On context switch, store CPU registers in TASK A's table.

2. Consult the ready list to find out which task is executed next.

3. Load new task's registers from its table.

4. Start executing TASK B.

Context switching

Special tables, often called task control blocks, store all the information the system requires about the task, for example, its memory usage, its priority level within the system and its error handling. It is this context information that is switched when one task is replaced by another.

The 'ready' list contains all the tasks and their status and is used by the scheduler to decide which task is allocated the next time slice. The scheduling algorithm determines the sequence and takes into account a task's priority and present status. If a task is waiting for an I/O call to complete, it will be held in limbo until the call is complete. Once a task is selected, the processor registers and status at the time of its last context switch are loaded back into the processor and the processor is started. The new task carries on as if nothing had happened until the next context switch takes place. This is the basic method behind all multitasking operating systems.

The M68000 architecture has extensive support for such operating systems: its SUPERVISOR/USER partitioning allows the kernel to run in SUPERVISOR mode with application tasks executing in the USER state. The timer interrupts can be given a specific interrupt vector number and handler, which will automatically switch the processor from running application tasks in its USER mode into SUPERVISOR mode to perform the context switch. The MOVEM provides a single instruction which will then load or reload the register set from a task control block in memory. This system partitioning prevents any exceptions caused by application tasks from corrupting the kernel, which could simply abort the erroneous task and carry on without it .

The 16 different TRAP instructions provide an elegant method of implementing operating system calls by causing an exception and allowing a handler, within the kernel, to perform the request. The single step facility through the TRACE exception facilitates system debuggers (many 8 bit single step routines required special external hardware) and so on.

Multitasking and communication support

Support for intertask communication via semaphores was made possible via the TAS instruction and its associated read-modify-write cycles. If a resource is to be shared between processors or tasks running under a multiprocessor operating system, any shared resource such as I/O, peripherals and memory must be controlled so that only one processor or task can access it at one time. This is essential to prevent task A from overwriting data that task B was using and thus cause system corruption. A semaphore is frequently used to control such access. Before accessing such resources, the associated semaphore is checked. If it is clear then no-one else is using it and access is permitted. The task must then set the semaphore to lock out other tasks. If it is set, the task must repeatedly check the semaphore until it is clear and the resource is available. This can be done by the test and set instruction, TAS. However, there is a potential problem: in between checking and setting the semaphore, another task could also check and set the semaphore with the result that both tasks now have access. This is highly undesirable ! This can occur if the read cycle, i.e testing the semaphore status, and the write cycle, i.e. the semaphore setting can be split from two consecutive cycles into two independent ones. This can occur if another processor requests the bus in the middle or if context switch occurs in a multiprocessing operating system. To prevent this, the M68000 family has an indivisable read-modify-write cycle. This cannot be interrupted and neatly prevents the problem. This cycle type is indicated by continued assertion of AS* through both the cycles.

High level language support

The M68000 provides substantial help for high level languages by providing many specific instructions and functions to reduce the amount of work that a compiler has to perform. Condition codes are set after the completion of virtually every instruction, thus removing many comparisons that might otherwise have to be made. Procedure calls are helped by the LINK and UNLINK instructions which allow data frames to be attached and removed from the stack with a single instruction. These allow nested subroutines to use the stack areas as temporary data storage without losing track or control.

The bit manipulation instructions are ideal for real-time applications, semaphore control, or any other application which needs to control data at the individual bit level. The alternative to this involves bit masking or a series of logical functions: both take multiple instructions to implement.

The biggest advantage for compliers is the large general purpose register set. The diagram shows the main advantage of general purpose registers: with a dedicated register set, the existing data must be moved out to other registers or to the external memory stack, the new data moved into its place and the finally the specific operation performed. This sequence must be repeated each time fresh data is involved. This needs multiple instructions and results in a considerable amount of time being spent simply moving data from one place to another without actually processing it. With a large number of variables as shown, this problem is even worse.

The data is stored on the stack and therefore any access must go to external memory with the performance impact associated with the slower memory access. The M68000 register set allows variables to be stored in registers, allowing easy and fast access. The M6800 code example which multiplied two 16 bit numbers shows the code expansion that occurs when dealing with data larger than the processor can handle. The M68000 family with its 32 bit wide data paths does not suffer from this problem and thus allowed the easy development of 32 bit software right from day one.

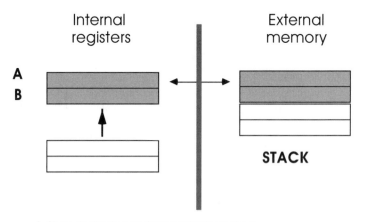

1. Store dedicated register contents on stack.
2. Move new values into dedicated registers.
3. Perform operation.

Data movement with dedicated registers

The concept of a large linear addressing space is of importance to compilers and other applications. The large space does not suffer from the problem and overhead incurred through segmentation. A segmented architecture like the Intel 80x86 splits the memory into individual 64 KByte segments which are controlled by a segment register. Each time an address crosses a segment boundary, software has to reload the segment register. While this did not seem to be a problem in 1979 when most applications would fit in a segment, it contributes to the overhead involved in using larger memory spaces. For compilers this simply restricts the direct addressing to only 64 Kbytes, which is reminiscent of the problems associated with bank switching techniques used with the older 8 bit generations. The M68000 family's ability to use large linear addressing has encouraged programmers to generate large applications with many facilities: the Apple Macintosh wordprocessor used to write this version of text uses 5 Mbytes for its application and 4 MBytes for its system software !

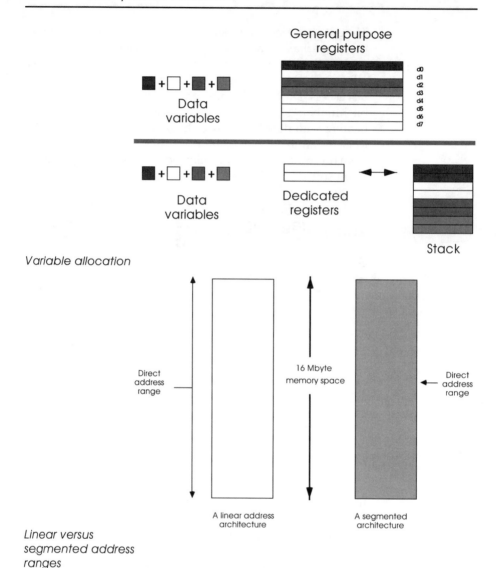

Variable allocation

*Linear versus
segmented address
ranges*

Start of a revolution

The architecture´s similarity to a mini computer, rather than its 8 bit predecessors, was now providing the necessary support for software developers to move operating systems, compilers and applications to the cheaper microprocessor environment and away from the discrete and more expensive mini and mainframe computers. This pull-down effect was to catalyse a rapid increase in their use and sounded ominous warnings to the computer manufacturing industry. However, the development of the basic M68000 architecture had another iteration to perform before the appearance of the full 32 bit implementations, the MC68020 and MC68030.

The MC68010 virtual memory processor

To replace mini and mainframes, the MC68000 needed to provide a more efficient mechanism to handle external memory faults, necessary for virtual memory operating systems such as UNIX, and provide a virtual machine capability. These two concepts came directly from mini and mainframe environments and were necessary to allow M68000 systems to become cost effective platforms for these types of applications. The MC68010 was a simple superset of the MC68000, with virtual memory and machine support built in. It appeared in 1983, four years after the MC68000, and had the same USER register set and ran MC68000 USER binary code without modification but faster by virtue of some microcode changes. From a hardware perspective, the parts were interchangeable and acted as plug-in replacements for each other.

Virtual memory support

A virtual memory operating system allows a program or task to access memory that is physically not present or is being used by another task. The technique usually involves splitting a task into smaller blocks, a memory management unit being used to translate each block's logical addresses into physical ones, some form of mass storage to act as a memory reservoir to store unused blocks and a protection system to protect all the blocks from corruption.

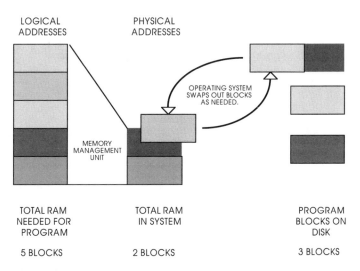

LOGICAL ADDRESSES

PHYSICAL ADDRESSES

OPERATING SYSTEM SWAPS OUT BLOCKS AS NEEDED.

MEMORY MANAGEMENT UNIT

TOTAL RAM NEEDED FOR PROGRAM

5 BLOCKS

TOTAL RAM IN SYSTEM

2 BLOCKS

PROGRAM BLOCKS ON DISK

3 BLOCKS

Using virtual memory to support large applications

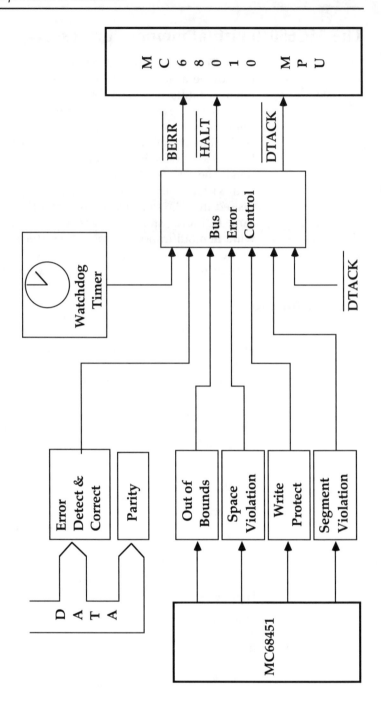

Bus error sources

There are usually two reasons behind using such a system: one occurs when the application requires more memory than is physically available and the second is when multiple tasks need to access separate memory but at the same address, as may happen within a multitasking operating system.

For this system to work, page faults must be detected and signalled back to the processor to halt the current bus cycle. Such faults occur when the processor tries to access non-existent or non-allocated memory, where an access goes across a page boundary or when the required block is resident on disk and needs to swapped with an unused block. In addition, bus faults may be generated by memory parity or EDC (error detection and correction) problems. These functions are easily performed by an MMU (memory management unit) like the MC68451 and some bus error routing logic as shown.

The real difficulty is what the processor should do once the page fault has been corrected and the cycle needs re-running. The MC68000 can be halted in mid-instruction but has to restart the interrupted instruction. This can cause a lot of challenges for the software handler to resolve: has the data within the processor been changed? If so, how? Has the external data been partially modified? If so, how can the instruction be restarted?

The MC68010 provides a hardware solution to solve this dilemma. When it receives a bus error, it switches into SUPERVISOR mode and starts to build an external stack frame using the SUPERVISOR stack pointer, A7´, just like the MC68000.

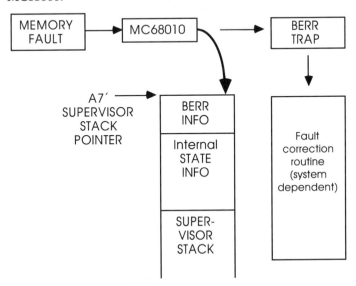

MC68010 bus error recovery - stack building

However, the frame is different and internal state information is added, in addition to the normal stack frame information containing the program counter, access address, etc. This state information is the contents of the processor's internal registers, latches, etc. which direct the execution of the instruction. This block is not defined or described and therefore is strictly 'off limits' to any manipulation or modification. The only publicly defined information within it is a processor mask number.

After the page fault has been corrected, an RTE (return from exception) instruction is executed, which reloads the processor with its previous contents to restore its state at the time of the error. Once finished, the MC68010 automatically completes the instruction execution by re-running the bus cycle that had previously failed. This time, no failure should be encountered.

There are a couple of interesting possibilities that this mechanism offers. First, *any MC68010* can recover this state information and continue the program that encountered the error. This allows freedom for any processor within a distributed multiprocessor system to complete an instruction that another has been unable to do. Secondly, it allows systems to use the RTE instruction with a predefined stack frame to set up the processor in an exact predetermined state. Both of these are valuable techniques for robust or fault tolerant systems.

Virtual machine support

A virtual machine environment allows several operating systems to execute concurrently on a single processor system, with each operating system thinking that it has total control of the system! In practice, one operating system is defined as the governing operating system, with the others running subservient to it. The governing operating system (GOS) executes in SUPERVISOR mode while all other virtual operating systems (VOS) and applications run in USER mode. Any requests for or attempts at system access by such applications or operating systems are detected by an M68000 privilege violation exception. This immediately invokes the GOS, which either rejects the request, performs it directly or performs it via the appropriate VOS.

To keep control of such a system requires a mechanism similar to that described for a multitasking operating system. Tables describing each operating system environment must be kept. When an exception occurs, the GOS always handles it initially. It first determines where the exception came from. If it was from an application or utility, the appropriate VOS will handle it. If it was from a VOS, possibly as a result of an application request, the GOS will service the request and pass the results to the VOS which will have appeared to have performed the service itself. Such interaction is simplified if different vector tables can be allocated to each operating system and selected as necessary. It would also be useful to be able to move data from SUPERVISOR to USER space to allow a VOS to receive data from the GOS, even if external hardware partitioning has been implemented.

These functions are provided via additional registers and instructions available to the SUPERVISOR.

MC68010 SUPERVISOR resource

Several additional registers have been added to the MC68010 SUPERVISOR programming model. They are the vector base register (VBR) and two alternative function code registers, one source and the other destination. Both of these registers can be loaded using the new MOVEC instruction. Loading the VBR with a value will offset the vector table from its original location at $000000 to anywhere in the memory map. This can provide a unique vector table for each VOS. The VBR is set to zero and is compatible with the MC68000 on power up.

31	16 15	0	
INTERRUPT STACK POINTER			A7'
	STATUS REGISTER		SR
VECTOR BASE REGISTER			VBR
SOURCE FUNCTION CODE			SFC
DESTINATION FUNCTION CODE			DFC

MC68010

The MC68010 SUPERVISOR register model

The new MOVES instruction will move data from one location to another but the contents of the alternate function code registers are output on the three function code pins instead of the normal code. With separate source and destination registers, data can be accessed from one space and transferred to another while maintaining the appropriate function codes necessary for correct external hardware decode.

To provide a completely secure SUPERVISOR environment, the MOVE from SR (status register) instruction was made privileged, whereas it was non-privileged with the MC68000. Although this instruction was hardly ever used, a small exception handler was necessary to ensure compatibility. It took the exception, identified the MOVE from SR command and then decided if the information could be passed to the USER.

Generic exception handlers were better supported by the inclusion of a vector number within the exception stack frame. A single handler could then be written to support multiple hardware resources within a system (e.g. one serial driver could control six serial ports). By checking the vector offset in the stack frame, the handler can immediately determine which port has interrupted it and use the appropriate addresses to access it directly.

Other improvements

The opportunity was taken with the MC68010 to re-evaluate many of the microcode sequences used internally to execute instructions. An example is the well known anomaly concerning the CLR instruction and the MC68000. It simply sets a memory location to zero, yet it actually performs an additional read cycle before writing zero into the location. In very fast code, a move immediate instruction is often used instead. This raises the question of why two instructions are needed to perform the same operation? They do have different effects on the condition codes but is that sufficient to warrant a separate instruction? The answer is probably not, however, it was present in the original M68000 design and to maintain software compatibility, it had to be included.

All 32 bit register based operations were speeded up, along with the shift and rotate functions and the arithmetic functions. The new divide algorithm was 25% faster that the MC68000 in performing that operation.

The most impressive improvement came with the implementation of 'Loop mode'. A block move of any size from anywhere to anywhere within the 16 Mbyte memory map can be performed by a simple two instruction loop as shown.

```
START   MOVE.L    $NO_WORDS,D0    ; move number of words
                                  ; into register D0
        MOVEA.L   $BLK1,A0        ; point A0 at start of BLK1
        MOVEA.L   $BLK2,A1        ; point A1 at start of BLK2
LOOP    MOVE.W    (A0)+,(A1)+     ; move word, move pointers
        DBEQ.L    D0,LOOP         ; decrement D0,if not zero
                                  ; goto LOOP
```

Example MC68000
block move

Examination of the MC68000 bus cycles show that each time the two-instruction LOOP is executed, both the MOVE and DBEQ instructions are fetched. This is not too inefficient, as the internal pipelines already have them loaded and decoded. The MC68010 detects this software structure, automatically locks the pipeline and will not repeatedly fetch the two instructions. This frees up the bus, allowing it to be dedicated to moving the data from one block to another, which is performed much faster. 'Loop mode' only works with loops involving the DBcc (decrement and branch on condition) instruction and another with no absolute or offset addressing and the decrement. Inserting another instruction, even an NOP, between the two will destroy the loop mode and force the repeated instruction fetches. This is often done in benchmarks and system testing to force the MC68010 to act like an MC68000.

While most of the improvements centred on software, the MC68010 bus interface did offer help to designers using fast memory systems. The bus error signal BERR* timing was relaxed so that it could be asserted after DTACK* and would be recognised by the processor, which would then ignore the original DTACK*. This allowed a designer to assume that every access would complete correctly and gain an extra half clock cycle for an error detection circuit to determine an erroneous cycle.

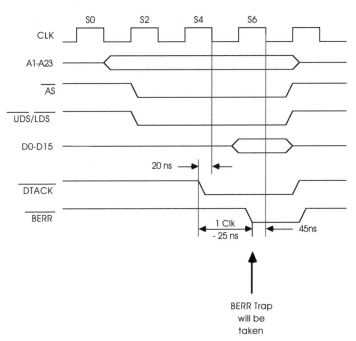

The MC68010 late bus error

The MC68008

The MC68008 also appeared at this time: it was an external 8 bit data version of the MC68000 in a 40 pin package. Its address bus was limited to 20 pins, giving a

1 Mbyte linear address space. It was aimed at low-cost systems and provides about 60% of the performance of the MC68000 at the same clock speed — not 50%, as even 16/32 bit machines need to fetch bytes occasionally!

Its main claim to fame was its inclusion in the Sinclair QL home computer, which was described as a 32 bit machine by its advertising material. This immediately created some controversy as it caused many arguments over exactly how a processor should be defined. Many people rang Motorola in the UK for adjudication. The answer was very simple but diplomatic: 'The MC68008 is an 8 bit external data bus microprocessor with a 16/32 bit internal architecture. Any 16 or 32 bit accesses are automatically performed as 2 or 4 sequential byte accesses and, therefore, software can use and specify 16 and 32 bit values.'

The story continues

Some observers have stated that the MC68010 is the processor that the MC68000 should have been first time around. For applications needing virtual memory etc. this may be a valid comment. It should be remembered though, that not all applications need virtual memory support etc., and for these, the MC68000 is an eminently suitable and cost effective solution due to its smaller die and cost.

CISC processors, like the M68000 family, have continued to provide more and more powerful instructions, addressing modes and hardware functions to reduce the amount of work required of the software running on it, as will be seen in the next chapter.

2   32 bit CISC processors

Almost as soon as the MC68000 and MC68010 had appeared, rumours started circulating about when the full 32 bit processor would be launched, what new features it would have and how much performance it would offer. The most common prediction was that it would be an MC68000 with simply full 32 bit data and address buses. It was argued that to offer any other features would take too many transistors, suffer from heat dissipation problems and just be too large a die to manufacture profitably.

This idea of simply expanding the buses appears on first inspection to be a very sensible stepping stone — it doubles the available data bus bandwidth and increases the linear address space. In reality, the performance increase it buys is not the factor of two, as one might expect. This is primarily due to the more frequent use of smaller data sizes — if software accesses bytes or words within a data structure, such an access is rarely sequential and is also fairly random. In such cases, fetching 32 bits and discarding 24 of them does not give any performance advantage. This is a similar situation, but in reverse, to that concerning the performance degradation suffered by a MC68008 mentioned in the previous chapter.

With the appearance of the MC68012, an MC68010 with a larger address bus, more fuel for the rumours of a simple upgrade was generated. When the MC68020 was announced in 1984, it surprised many in terms of its complexity, power and performance!

Enter HCMOS technology

With all microprocessor designs, it is the fabrication technology which will ultimately limit or facilitate new design ideas. This theme will recur throughout this book.

The main catalyst for the REDWOOD design team, which was responsible for the new 32 bit M68000 processor, was Motorola´s decision to move away from NMOS technology to high speed complementary metal oxide semiconductors (HCMOS). The new technology offered smaller transistor geometries, smaller power dissipation and higher packing density. This change give the designers about 200,000 transistors to play with almost a three times increase over the MC68000 — and allowed additional functions to be included.

The HCMOS process had another benefit — it was capable of supporting far higher clock speeds and allowed the processor to be built to 16 MHz design rules. This was twice the speed of the then standard 8 MHz MC68000 and MC68010. It was realised that with a processor running at twice the current system speeds, major problems would be created with memory and there would be a great danger of designing a very fast processor whose performance was restricted by memory design. Undoubtedly, memory systems would catch up but would this hold back the processor's acceptance and development?

Architectural challenges

The main challenge for CISC architectures is to reduce the number of clocks required to execute instructions. The M68000 family took the idea of splitting the operation into several stages within a single pipeline. Unfortunately, the architecture has features which decrease the pipeline efficiencies.

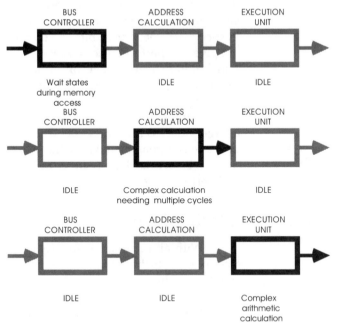

*Example of pipeline
stalls and delays*

Instruction op codes with additional operands requiring multiple memory cycles or wait states inserted in the memory accesses can delay the transfer within the pipeline and cause stalls. Complex address calculations can cause similar delays further into the pipeline. Conflict between data and instruction accesses on a single bus and coprocessor communication also cause the same result — pipeline delays. The diagram shows some examples using a simple three-stage pipeline.

Delays caused through effective address calculations and complex arithmetic operations could be solved by providing each pipeline stage with its own arithmetic logic unit (ALU) and using the faster clock speed. By not having to share an ALU between stages, calculation delays could be reduced and the faster clock obviously reduced the time needed to perform such operations. Analysis showed the likely bottleneck to be the memory subsystem.

CISC processors have a memory to memory architecture — they allow direct manipulation of data in external memory without having to explicitly use internal registers. This has the benefit of helping compiler writers at the expense of multiple bus cycles — while the bus controller is busy completing a data access, the rest of the pipeline is stalled when it could be executing other instructions. What was needed was another entry point into the pipeline, allowing instructions to be inserted from a local cache store.

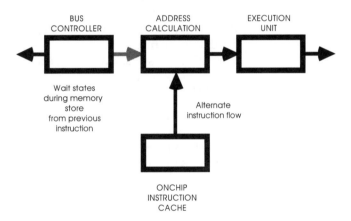

Feeding a stalled pipeline from an onchip cache

While an instruction is moving data to external memory as the last stage of its execution, the next instruction can be loaded into the pipeline. It can then be executed using the otherwise idle stages of the pipeline as shown. This allows the pipeline activities to be folded and the resultant instruction overlap gives the 'magical' result of zero clock cycle execution and an infinite MIPS performance! In reality, the execution time is hidden rather than reduced to nothing.

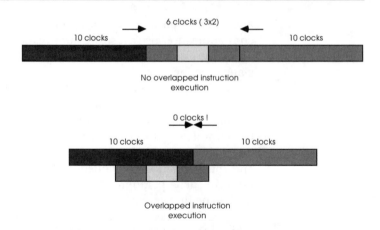

*Overlapping instruction
execution*

This was just one of the many revolutionary ideas implemented in the MC68020.

The MC68020 32 bit performance standard

The MC68020 was launched in April 1984 as the '32 bit performance standard' and in those days its performance was simply staggering — 8 million instructions per second peak with 2–3 million sustained when running at 16 MHz clock speed. It was a true 32 bit processor with 32 bit wide external data and address buses as shown. It supported all the features and functions of the MC68000 and MC68010, and it executed M68000 USER binary code without modification (but faster!).

- Virtual memory and instruction continuation were supported.
- The bus and control signals were similar to that of its M68000 predecessors, offering an asynchronous memory interface but with a three–cycle operation (instead of four) and dynamic bus sizing.
- Additional coprocessors could be added to provide such facilities as floating point arithmetic and memory management, which used this bus to provide a sophisticated communications interface.
- The instruction set was enhanced with more data types, addressing modes and instructions.
- Bit field data and its manipulation was supported, along with packed and unpacked BCD (binary coded decimal) formats.

An instruction cache and a barrel shifter to perform high speed shift operations were incorporated onchip to provide support for these functions.

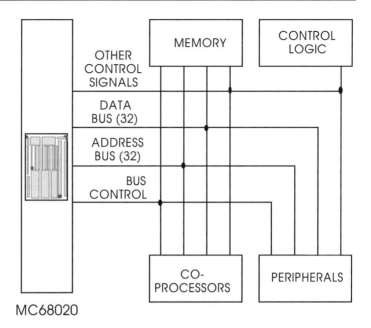

MC68020

*A simple MC68020
system*

The device was truly state of the art — its die contained 200,000 transistors fabricated in a 2.25 micron HCMOS technology, packaged in a 114 pin PGA (pin grid array) and typically dissipated 750 mW, the same as an MC68000. The original design was breadboarded using 14 boards, each containing 300 MSI (medium scale integration) devices and tested with over 500,000 lines of diagnostic software.

The internal design philosophy

The internal block diagram shows the basic philosophy behind the design. The device has three separate ALUs (arithmetic logic units).

- One unit performs effective address calculations.
- A second unit perfomsr operand calculations.
- A third unit performs arithmetic instructions.

These three unit prevent pipelines stalls due to oversubscribed resources.

The bus controller and sequencer can now operate independently of each other, allowing simultaneous data and instruction fetches. While data is being moved on the external bus, the next instruction can be fetched from the instruction cache and possibly executed. Again, this parallelism makes the best use of the design and improves performance.

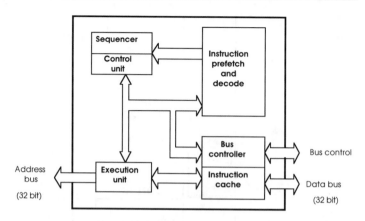

*The MC68020 internal
block diagram*

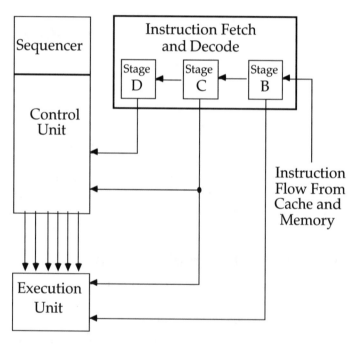

*The MC68020 internal
pipelining*

The actual pipeline used within the design is quite
sophisticated. It is a four–stage pipe with stage A consisting of
an instruction router which accepts data from either the
external bus controller or the internal cache. As the instruction
is processed down the pipeline, the intermediate data can
either cause micro and nanocode sequences to be generated to
control the execution unit or, in the case of simpler instructions, the data itself can be passed directly into the execution
unit with the subsequent speed improvements.

The programmer´s model

The programmer´s USER model is exactly the same as for the MC68000, MC68010 and MC68008. It has the same eight data and eight address 32 bit register organisation. The SUPERVISOR mode is a superset of its predecessors. It has all the registers found in its predecessors plus another three. Two registers are associated with controlling the instruction cache, while the third provides the master stack pointer.

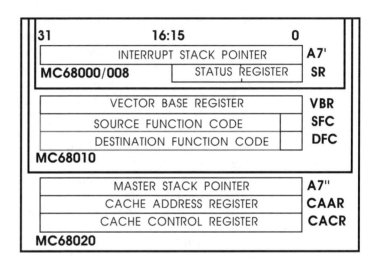

31	16:15	0	
INTERRUPT STACK POINTER			**A7'**
MC68000/008	STATUS REGISTER		**SR**
VECTOR BASE REGISTER			**VBR**
SOURCE FUNCTION CODE			**SFC**
DESTINATION FUNCTION CODE			**DFC**
MC68010			
MASTER STACK POINTER			**A7''**
CACHE ADDRESS REGISTER			**CAAR**
CACHE CONTROL REGISTER			**CACR**
MC68020			

The M68020 SUPERVISOR programming model

The supervisor uses either its master stack pointer or interrupt stack pointer, depending on the exception cause and the status of the M bit in the status register. If this bit is clear, all stack operations default to the A7´ stack pointer. If it is set, interrupt stack frames are stored using the interrupt stack pointer while other operations use the master pointer. This effectively allows the system to maintain two separate stacks. While primarily for operating system support, this extra register can be used for high reliability designs.

The MC68020 instruction set is a superset of the MC68000/MC68010 sets. The main difference is the inclusion of floating point and coprocessor instructions, together with a set to manipulate bit field data. The instructions to cause a trap on condition, perform a compare and swap operation and a 'call-return from module' structure were also included. Other differences were the addition of 32 bit displacements for the LINK and Bcc (branch on condition) instructions, full

32 bit arithmetic with 32 or 64 bit results as appropriate and extended bounds checking for the CHK (check) and CMP (compare) instructions.

The bit field instructions were included to provide additional support for applications where data does not conveniently fall into a byte organisation. Telecomms and graphics both manipulate data in odd sizes — serial data can often be 5, 6 or 7 bits in size and graphics pixels (i.e. each individual dot that makes a picture on a display) vary in size, depending on how many colours, grey scales or attributes are being depicted. The MC68020 bit field data type is from 1 to 32 bits in size. It is defined by three terms — a starting address anywhere within the 4 Gigabyte linear addressing space, a starting bit which can be defined anywhere in a 4 Gigabit ($\pm 2^{31}$ range) and a data width which determines the size. The manipulation instructions allow data to be inserted, extended, tested, cleared, set or changed. The first set bit within a field can also be detected.

The MC68000 and MC68010 can perform these tasks by simulating instructions with software routines involving the standard logical shift and bit test instructions. However, the execution time is often dependent on both the number of bits involved and the number of shifts needed. Examination of the instruction timings shows that a logical shift instruction execution time is $(10 + 2n)$ clocks where n is the number of bits. To reduce this overhead, the MC68020 has a hardware accelerator called a barrel shifter, which can perform a shift in a single clock, irrespective of the number of bits shifted. This hardware support is virtually essential for any bit field operations. It is also used when packing and unpacking BCD data.

Dn	Data Register Direct
An	Address Register Direct
(An)+	Address Reg. Indirect w/ Post-Increment
-(An)	Address Reg. Indirect w/ Pre-Decrement
d(An)	Displaced Address Register Indirect
d(An,Rx)	Indexed, Displaced Address Reg.
d(PC)	Program Counter Relative
d(PC,Rx)	Indexed Program Counter Relative
#xxxxxxxx	Immediate
$xxxx	Absolute Short
$xxxxxxxx	Absolute Long

MC68000/008/010

(bd,An,Xn.SIZE*SCALE)	Register Indirect
	Memory Indirect
([bd,An,Xn.SIZE*SCALE],od)	Pre-Indexed
([bd,An],Xn.SIZE*SCALE,od)	Post-Indexed
	Program Counter Memory Indirect
([bd,PC,Xn.SIZE*SCALE],od)	Pre-Indexed
([bd,PC],Xn.SIZE*SCALE,od)	Post-Indexed

MC68020/MC68030

The MC68020 addressing modes

The addressing modes were extended from the basic M68000 modes, with memory indirection and scaling. Memory indirection allowed the contents of a memory location to be used within an effective address calculation rather than its absolute address. The scaling was a simple multiplier value 1, 2, 4 or 8 in magnitude, which multiplied (scaled) an index register. This allowed large data elements within data structures to be easily accessed without having to perform the scaling calculations prior to the access. These new modes were so complex that even the differentiation between data and address registers was greatly reduced: with the MC68020, it is possible to use data registers as additional address registers. In practice, there are over 50 variations available to the programmer to apply to the 16 registers.

The new CAS and CAS2 'compare and swap' instructions provided an elegant solution to linked list updating within a multiprocessor system. A linked list is a series of data lists linked together by storing the address of the next list in the chain in the preceding chain. To add or delete a list simply involves modifying these addresses. In a multiprocessor system, this modification procedure must occur uninterrupted to prevent corruption. The CAS and CAS2 instruction meets this specification. The current pointer to the next list is read and stored in Dn. The new value is calculated and stored in Dm. The CAS instruction is then executed. The current pointer value is read and compared with Dn. If they are the same, no other updating by another processor has happened and Dm is written out to update the list. If they do not match, the value is copied into Dn, ready for a repeat run. This sequence is performed using an indivisible read-modify-write cycle. The condition codes are updated during each stage. The CAS2 instruction performs a similar function but with two sets of values. This instruction is also performed as a series of indivisible cycles but with different addresses appearing during the execution.

Bus interfaces

Many of the signals shown in the pin out diagram are the same as those of the MC68000 — the function codes FC0-2, interrupt pins IPL0-2 and the bus request pins, RESET*, HALT* and BERR* perform the same functions.

With the disappearance of the M6800 style interface, separate signals are used to indicate an autovectored interrupt. The AVEC* signal is used for this function and can be permanently asserted if only autovectored interrupts are required. The IPEND signal indicates when an interrupt has been internally recognised and awaits an acknowledgement cycle. RMC* indicates an indivisible read-modify-write cycle instead of simply leaving AS* asserted between the bus cycles. The address strobe is always released at the end of a cycle. ECS* and OCS* provide an early warning of an impending bus cycle, and indicate when valid address information is present on the bus prior to validation by the address strobe.

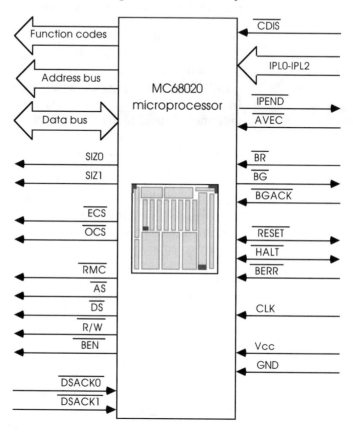

The MC68020 device pin-out

The M68000 upper and lower data strobes have been replaced by A0 and the two size pins, SIZE0 and SIZE1. These indicate the amount of data left to transfer on the current bus cycle and, when used with address bits A0 and A1, can provide decode information so that the correct bytes within the 4 byte wide data bus can be enabled. The old DTACK* signal has been replaced by two new ones, DSACK0* and DSACK1*. They provide the old DTACK* function of indicating a successful bus cycle and are used in the dynamic bus sizing. The bus interface is asynchronous and similar to the M68000 cycle but with a shorter three–cycle sequence, as shown.

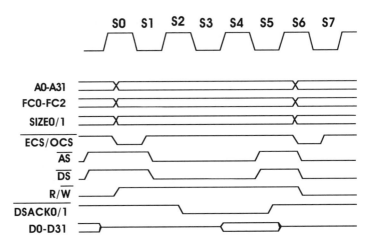

MC68020 bus cycle timings

Dynamic bus sizing

Most of the systems available at the MC68020's introduction were either 8 or 16 bits in size. Even VMEbus designs were based around the 16 bit format, although today it is widely accepted and used in its full 32 bit format. The problem comes with trying to combine 8, 16, and 32 bit data paths. The processor can support these types explicitly within its instruction set so software could be forced to use certain data sizes with certain address ranges. The first 1 Mbyte could be 32 bits wide, the next 16 Mbytes be 16 bits and so on. This does, however, constrain software development — compilers cannot use the a single 32 bit move instruction to fetch a 32 bit wide variable without performing some address range check. Equally, if the hardware design was changed, how would this affect software? One solution could be to adopt a standard 32 bit data path, but this would be expensive considering past design investment. While today this is totally feasible, 4–5 years ago, there were simply not the products off the shelf!

Dynamic bus sizing allows the processor to change the data bus width on a cycle by cycle basis without prior knowledge of the memory port width. The processor always attempts the largest transfer it can (32 bits) and indicates the number of bytes needed to transfer by coding the two SIZE signals. The memory port takes as much data as possible and indicates a success and the number of bytes taken by responding with the two DSACK signals. The processor sees this, calculates how many bytes are left to transfer, shuffles the data up on its bus and automatically starts another bus cycle.

Again, the port replies and the process repeats until all the data is transferred. This system means that software can specify whatever data size it needs and, providing the memory interface observes this protocol, it can fetch this data from any size or combination of external data widths. The only penalties are the extra bus cycles and increased bus activity.

This flexibility is further enhanced by the processor's ability to fetch data that lies across word and long word boundaries. Such misaligned data is fetched as a series of aligned accesses and reconstituted internally. Again, the only penalties are extra bus cycles and increased bus activity. Instructions, however, are always fetched on a word boundary and any attempt to fetch a misaligned instruction results in an exception.

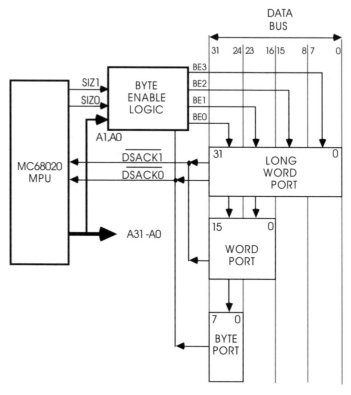

DSACK1	DSACK0	MEANING
HI	HI	Insert wait state
HI	LO	Complete cycle, port size = 8 bits
LO	HI	Complete cycle, port size = 16 bits
LO	LO	Complete cycle, port size = 32 bits

MC68020 dynamic bus sizing

On chip instruction cache

One of the perennial problems faced by any fast processor designer is that of slow memory accesses. Often a processor can execute instructions far faster than the system can provide them or any associated data. This von Neumann bottleneck, named after the pioneering mathematician, is caused by delays in accessing external memory, such as wait states, bus arbitration, etc. The MC68020 doubled the processor clock speed to 16 MHz overnight and reduced an MC68000 bus cycle by 25% from 4 clocks to 3. By today's standards, these speeds are slow, but in 1984 this was state-of-the-art and presented a major obstacle. One way of solving this problem is to use cache memory. Here, instructions and/or data can be stored in local very fast memory. The first access goes to main memory and is slow, but a copy is stored in the fast cache memory. Any subsequent access to this location is now made to the cache, giving a faster access and better performance. Providing software is written as a series of tight instruction loops, some benefit can be obtained. Cache designs and their characteristics are becoming more and more important for processor architectures and will be more fully explored in later chapters.

The MC68020's internal 256 byte instruction cache was added to prevent the degradation expected within a system with a poor memory performance. It holds about 100 instructions and is organised as a 64 by 4 byte array. Accessing instructions from the cache can occur simultaneously with an external data fetch and this reduces the bus bandwidth needed by the processor. Cache accesses take only two cycles, giving a minimum 33% improvement over external fetches. Typically, an MC68020 uses 60% bus bandwidth compared with the 80–90% of an MC68000. By being instruction only, there are no associated stale data problems. From a programming viewpoint, it is very simple — there are two programming registers available to the SUPERVISOR which can enable/disable the cache or freeze its contents. Normally, any cache updating is performed automatically using a special replacement algorithm. This determines which cache entry will be used when data is fetched from external memory. When the cache is frozen, no updating is performed. In addition, either the whole cache or specific entries can be cleared. The beauty of the MC68020 cache is its simplicity — the cache is enabled and software executes typically 15–20% faster and the maintenance and updating are performed internally and automatically. The only maintenance required is when the cache is used with memory management within a multitasking operating system. The cache appears on the logical side of the memory management unit and does not automatically know when a context switch has occurred. When a switch does take place, the cache contents are no longer valid and must be cleared. With the MC68020, this can be simply performed by an instruction within the operating system kernel.

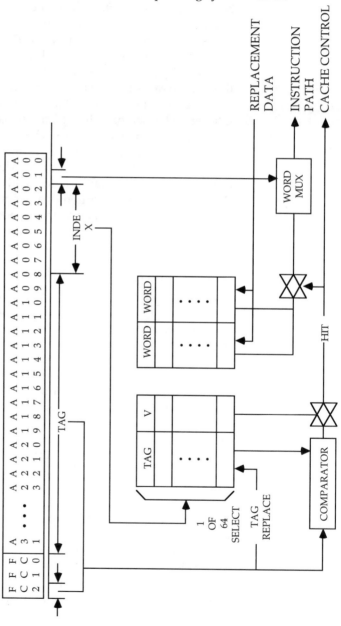

The MC68020 internal instruction cache

The cache system works very simply. When an instruction fetch is started, the address is supplied to the bus controller and to the cache. If the data is present in the cache, a cache hit signal prevents the bus controller from continuing the external bus cycle. (Strictly speaking, the cycle is never really started because the address strobe, AS*, is not asserted.) If the data is not cached, a miss occurs, the external bus cycle continues and the instruction is fetched from external memory. A copy of the instruction is then used to update the internal cache. By supplying the address to both the bus controller and cache, the external bus cycle is not delayed if there is a cache miss. Alternatively, if the bus controller cannot start the external cycle due to prior cycles or bus arbitration, the cache lookup is not delayed.

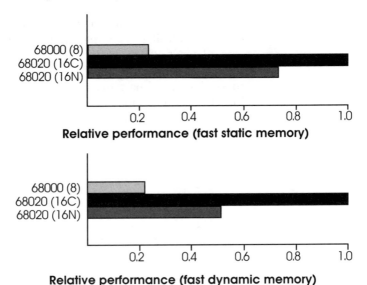

Relative performance (fast static memory)

Relative performance (fast dynamic memory)

BASED ON A RESEARCH STUDY BY SPERRY PUBLISHED IN IEEE MICRO
AUGUST 1986
(NOTE: C = on-chip cache enabled, N = on-chip cache
disabled)

*MC68020 performance
figures*

The lookup works by taking bits 2 to 8 of the access address and using them to index into the 64 element tag array. The remaining higher address bits, together with function code 2 to differentiate between SUPERVISOR and USER, are then compared with the tag memory contents. If they match, there is a cache hit and address bit 1 is used to select which of the two words stored in the cache data memory is sent via the instruction path. If there is a mismatch, or the tag is marked as not valid, a cache miss is caused. When the instruction is brought back from external memory, it and its associated tag entry replace the original cache entry. If the cache has been frozen, no updating occurs. These mechanisms are performed automatically.

The cache obviously provides great benefit as can be seen from the histogram. The Sperry (now Unisys) benchmarks were based around the EDN benchmark suite which tests interrupt response, bit manipulation, string sorting, etc. With fast static RAM, the improvement is about 25–30%. With slower dynamic memory, this figure is increased to 50%. This performance gain is obtained by simply executing one instruction to enable the cache.

The cache causes difficulties when trying to work out how many clocks a particular instruction will take. As shown, it is possible for a register–based instruction to be executed in zero clock cycles when it can be fetched from the cache while the bus controller is completing the preceding data movement. The NOP instruction was enhanced to provide a software method of synchronising the execution pipeline and bus controller, and thus prevent overlapping. This infinite performance is not achieved every time and is dependent on preceding instruction flow, bus activity and the state of the cache.

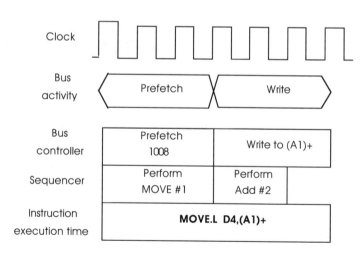

Address	Op-code	Code sequence	Effective Execution time
1000	22C4	MOVE.L D4,(A1)+	6 Clock cycles
1002	DA84	ADD.L D4,D5	*0 Clock Cycles*

Instruction overlapping

This is extremely difficult to predict without actual measurement, although the instruction timings stated in the MC68020 user's manual are divided into three different types:

- best case with maximum overlap,
- cache case, and
- worst case.

In reality, the actual performance lies somewhere around the cache case figure, but it is totally software and system dependent.

While software engineers can exploit the cache to its full by freezing the cache contents with time–critical routines, these techniques cause great difficulties for their hardware colleagues. Such small loops, including diagnostic test loops, can easily be completely cached. This removes any external instruction fetches and the processor only uses the bus for data reads and writes. If the data is register based, the processor sits there apparently doing nothing. In addition, unlike its predecessors, the MC68020 does not halt processing when the external HALT signal is asserted. It carries on executing instructions from its cache until an external fetch is needed. The halt signal is often used to provide external cycle–by–cycle single stepping but this is no longer so useful unless the cache is disabled. Further confusion is caused by the external bus cycles only providing an intermittent view of the processing going on internally. Additional support was needed to solve these problems.

Debugging support

The most important feature of debugging is the CDIS pin which disables the internal cache when asserted. With it asserted, all bus cycles take place externally, thus allowing analysers and emulators access. The cache does cause some interesting phenomena, however. The normal method of breakpointing code is to replace an instruction with an illegal one to force an exception when it is executed. Alternatively, hardware can analyse the bus cycle traffic until the breakpoint address is accessed. At this point, the hardware can take over and stop the processor. Neither of these techniques works very well with the MC68020. If the instruction is replaced after it has been cached, the cached instruction will be fetched in preference, resulting in a breakpoint that appears never to be reached. The cache must either be cleared or disabled, hence the CDIS pin.

The hardware method is fine, except that it cannot determine exactly when an instruction is executed within the MC68020. In a small loop, with a breakpoint set on the first instruction after the loop, the breakpoint address is prefetched every time the loop is executed and therefore the hardware stops the processor on every loop. Again, this is not very useful.

Most emulators use a combination of the two techniques. The address traffic is analysed and when a breakpoint address appears, the hardware jams an illegal instruction on the data bus. If this is a simple prefetch, the illegal instruction is not be executed and nothing happens. If the access is cached, the next access fetches the illegal instruction correctly from the cache. If the instruction is then executed, this forces an exception and stops the processor as required. Another advantage of this mechanism is its ability to set breakpoints in read only memory.

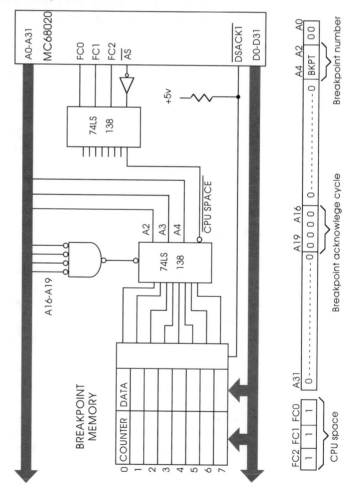

The MC68020 break-point instruction

The MC68020 design provided additional support by way of the breakpoint instruction. This instruction forces a special breakpoint CPU space cycle to take place and allows external hardware to supply the option of forcing the exception or providing the original instruction, thus allowing execution to continue. This process needs external hardware support and is therefore totally system dependent.

The original T trace bit of the MC68000 was also enhanced. Its function is performed by a T1 trace bit within the status register which, when set by the supervisor, takes the trace exception after the completion of any instruction. If the T0 bit is set instead (it is not allowed to set both bits), the trace exception is taken only after a change of flow. Such changes in flow can be caused by branching and jumping. The trace facility is best looked at as an extension of the basic instruction. If tracing is enabled, an instruction is not deemed to have completed until any trace processing has been completed. Only if an address or bus error occurs during the instruction execution will the trace processing be deferred. When the exception is taken, the trace bits are cleared, preventing any further tracing.

Coprocessor interface

Even with the large number of transistors available, there were not enough to provide memory management or floating point support onchip. The obvious solution was to use additional chips with an interface as transparent as possible. For the programmer, this means simply bringing additional registers and instructions to the software model. For the hardware designer, it should be a simple interface using standard signals and should also allow these coprocessors to be optional. The MC68020 coprocessor interface uses no special signals, supports both logical and physical caches and allows a coprocessor to act as a bus master in its own right. The basis for the mechanism are the F line exceptions and standard bus cycles.

The M68000 family treated the F and A line instructions in a special way. Their execution causes an exception and allows software to emulate their action. These instructions are used on the MC68020 to provide coprocessor instructions for floating point arithmetic and memory management facilities. When such an instruction is executed, the MC68020 starts a CPU space cycle. This is signified by the three function code outputs all going high. For the 16 bit family members, this indicated an interrupt acknowledgement cycle only. A CPU space cycle can still be an interrupt acknowledgment but can also indicate a coprocessor cycle, a breakpoint cycle or an access level control cycle. These different types are indicated by different bit pattern within the CPU space type field, bits 16–19 of the address bus. For a coprocessor cycle, this pattern is 0100 with bits 13–15 identifying the coprocessor identity number and bits 0–4 , the register number.

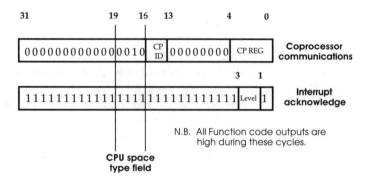

Different CPU space cycles

The protocol allows up to eight coprocessors to be added. Usually only two sockets are used, with the MC68851 memory management unit always being socket 0, together with a MC68881 or MC68882 floating point unit in socket 1. The extra registers are located on the coprocessor and the processor acts as an intelligent DMA (direct memory access) controller, fetching data as needed. The processor can be stalled or released to allow either sequential or concurrent operation as needed.

During the start of coprocessor instruction execution, the CPU sends the op code to the coprocessor using a coprocessor communications CPU space cycle. The coprocessor id is decoded and used to physically select the correct coprocessor which then takes the op code and sends a response back to the CPU. This response could indicate that the coprocessor is busy and cannot execute the op code presently, that more data or operands are needed or that no further service is needed. If the coprocessor is busy, the CPU repeatedly tries to send the op code until it is successful or until an interrupt or other exception occurs. If the coprocessor is not capable of fetching data from memory itself, it uses the CPU as a DMA controller to perform such tasks. The coprocessor can force sequential execution by reserving the CPU for such requests and not actually requesting any. Once the CPU has finished servicing the coprocessor, it is free to execute the next instruction.

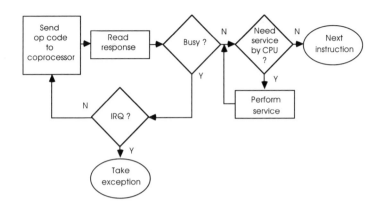

*The MC68020
coprocessor Inter-
face(simplified)*

If the coprocessor instruction is invalid or the coprocessor hardware is not present, the normal exception is taken and software is called to reconcile the problem. This coprocessor interface can be simulated using standard M68000 bus cycles with additional logic and software so that they can be used as peripherals with the other family members.

MC68881 and MC68882 floating point coprocessors

The MC68881 floating point coprocessor provides hardware support for calculations using the full IEEE Floating Point Standard (P754 Rev 10) and transcendental functions (sin, tan, etc.). All calculations are performed to 80 bit precision and many mathematical constants are stored internally in ROM for immediate use. It adds eight 80-bit floating point registers, FP0-FP7, with three control registers to the USER programming model.

The 68 pin PGA device acts as a non-DMA coprocessor (it relies entirely on the MC68020 for its data supply) and can be used as a peripheral with MC68000/010/008 processors. The interface is very simple, requiring only simple logic to decode the processor ID and register selection from the address bus during a CPU space coprocessor cycle to generate a chip select.

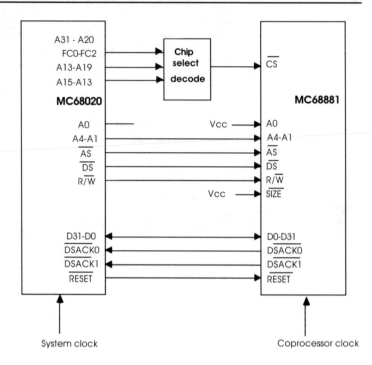

*A MC68020/MC68881
block diagram*

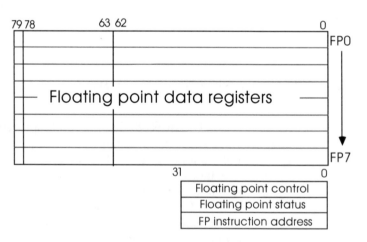

*MC68881/882 register
set*

Its performance was unsurpassed when it first appeared, although by today´s standards, it is not a performance leader. However, considering the faster clock speeds and its wide set of functions, it still offers a cheap and easy way of increasing performance. In extended precision, it can typically perform an addition in 2.2 μs, a multiply in 3.2 μs and a divide in 4.6 μs at a coprocessor clock of 25 MHz.

The MC68882 is a 'go-faster' version of the MC68881: it is upward and downward software compatible with MC68881, is hardware pin compatible and available in faster clock speeds (e.g. 33 and 40 MHz). It was designed to complement the MC68030 but is equally suitable for MC68020 designs. The performance is gained by increasing the onchip concurrency. The operand pre-fetch, data conversion and computation is all pipelined, with additional hardware to expedite operand loads and stores. The floating point registers are dual ported to allow simultaneous loading and computation.

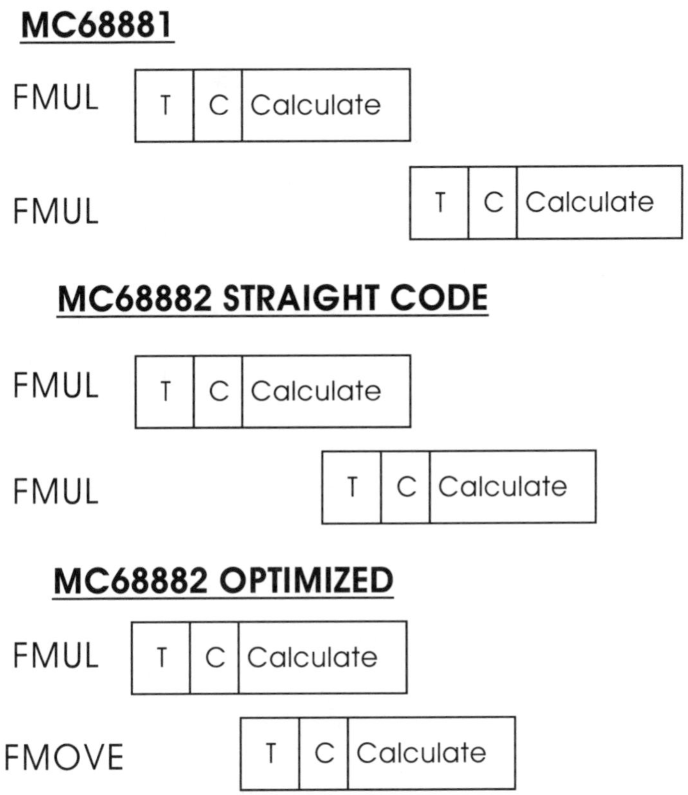

MC68881

FMUL | T | C | Calculate |

FMUL | T | C | Calculate |

MC68882 STRAIGHT CODE

FMUL | T | C | Calculate |

FMUL | T | C | Calculate |

MC68882 OPTIMIZED

FMUL | T | C | Calculate |

FMOVE | T | C | Calculate |

n.b. T = transfer, C = conversion.

MC68882 optimisation

The diagram above shows how the MC68882 achieves its optimisation. If it executes standard MC68881 code, it can overlap the transfer and conversion parts of the instruction processing so that they can commence during the calculation phase, instead of waiting for its completion. This will typically improve performance by a factor of about 1.4. Further optimisation can be obtained by alternating calculations with data moves where complete overlap occurs, giving a 1.7 to 2 times improvement. Both these figures are totally software dependant, however. It is possible, in certain cases, for far higher increases.

The MC68851 paged memory management unit (PMMU)

The MC68851 paged memory management unit is an intelligent coprocessor capable of external bus accesses. It provides full support for systems that need memory management to allow multiple tasks to use the same or overlapping sets of logical addresses. The PMMU sits between the CPU and memory.

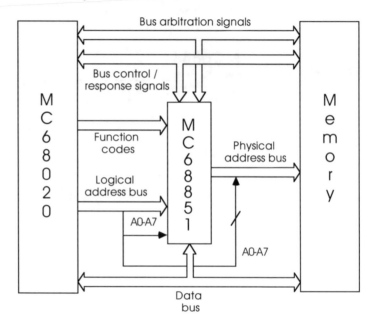

Connecting an MC68851 to a MC68020

Such sets can be declared as read only, SUPERVISOR only, etc., so that they can be protected. The device supports full 32 bit logical and physical addresses without multiplexing the buses. Address translations are performed in 45 ns due to an onchip address translation cache (ATC) which holds the 64 most recent translations so that they are immediately available. This reduces the translation delay to a single cycle. When the translation entry is not held in the cache, the PMMU acts as a bus master and performs a walk through its translation tables to get the correct translation. These tables are stored in external memory. The page sizes are programmable from 256 bytes to 32 Kbytes.

The address translation mechanism uses a root pointer to select a translation table. Part of the logical address, the table index A (TIA) is used to index into this table. The table contents are used with the lower logical address bits to form the physical address. The MC68851 supports multiple table levels and root pointers which allow large numbers of pages to be allocated without very large tables. Memory management techniques are more fully described later.

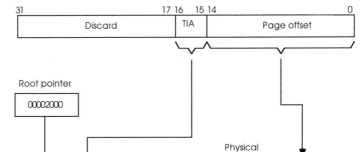

Logical Address = $00013526:

The MC68851 address
translation mechanism

In retrospect, the MC68851 was too sophisticated for many applications and offered facilities that were never used. It has even been suggested that over 45% of the chip´s hardware is never used. The real problem with the device (or any external memory management unit) is tracking processor speed increases. The fastest version only supports 20 MHz operation, against 33 MHz for the MC68020. With the appearance of the MC68030 together with its integrated MMU and faster clock speeds, the MC68851/MC68020 lost much of its appeal.

The MC68030: the first commercial 50 MHz processor

The MC68030 appeared some 2–3 years after the MC68020 and used the advantage of increased silicon real estate to integrate more functions on to a MC68020-based design. The differences between the MC68020 and the MC68030 are not radical — the newer design can be referred to as evolutionary rather than a quantum leap. The device is fully MC68020 compatible with its full instruction set, addressing modes and 32 bit wide register set. The initial clock frequency was designed to 20 MHz, some 4 MHz faster than the MC68020, and this has yielded commercially available parts running at 50 MHz. The transistor count has increased to about 300,000 but with its smaller geometries, die size and heat dissipation are similar.

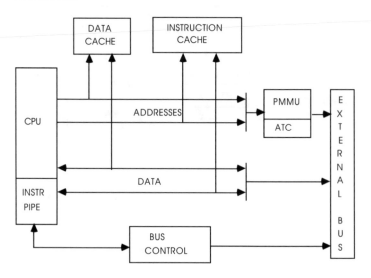

The MC68030 internal block diagram

Memory management has now been brought onchip with the MC68030 using a subset of the MC68851 PMMU with a smaller 22 entry on chip address translation cache. The 256 byte instruction cache of the MC68020 is still present and has been augmented with a 256 byte data cache.

Both these caches are logical and are organised differently from the 64 x 4 68020 scheme. A 16 x 16 organisation has been adopted to allow a new synchronous bus to burst fill cache lines. The cache lookup and address translations occur in parallel to improve performance.

The processor supports both the coprocessor interface and the MC68020 asynchronous bus with its dynamic bus sizing and misalignment support. However, it has an alternative synchronous bus interface which supports a two–clock access with optional single-cycle bursting. The bus interface choice can be made dynamically by external hardware.

Both cycles start with the address and other information appearing on the bus and being qualified by the address strobe. Whereas the asynchronous cycle is terminated by the DSACK signals, the hardware asserts STERM* to indicate that a synchronous transfer is taking place. If only a single transfer is needed, the data is latched on the second clock and STERM* is negated. If the processor read operation is filling an internal cache line (indicated by the CBREQ* cache burst request line) and the memory is capable of generating the next three addresses, a burst fill transfer can be accomplished. Here, the memory interface performs the normal synchronous access but does not negate STERM*.

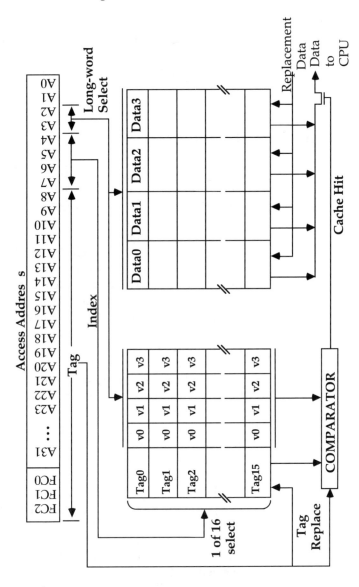

The MC68030 internal cache organisation

The processor supplies the next three locations on successive clock edges for the page or nibble mode memories to accept. The main advantage of this system is in reducing the necessity for fast static RAMs to realise the processor's performance. With 120 ns page mode DRAM, a 4-1-1-1 (four clocks for the first access, single cycle for the remaining burst) memory system can easily be built. Each 128 bits of data fetched in such a way take only 7 clocks compared with 5 in a no wait state system. If bursting was not supported, the same access would take 16 clocks. This translates to a very effective price performance — a 4-1-1-1 DRAM system gives about 90% of the performance of a more expensive 2-1-1-1 static RAM design.

Memory management

The MC68030 MMU supports 32-Bit logical and physical addresses with write and supervisor protection. It has multiple level tables and performs its external table walks under hardware control. The onchip 22 entry ATC is used to store the most frequently used translation descriptors. The device has an extra cache inhibit bit which allows the memory management software to declare memory areas non-cacheable. Accesses to these areas assert a cache inhibit signal externally, allowing the processor to tell external memory caches not to cache the data. All MMU translations are performed in parallel with the cache access and therefore, unlike the MC68851 which inserts a wait state, can perform address translations with no additional overhead.

Eight page sizes are supported ranging from 256 bytes to 32 Kbytes in size and two root pointers are allocated to supervisor and user accesses.

To prevent system ATC entries being overwritten by applications and thus causing additional delays while performing system calls, two transparent windows have been provided. These allow two memory areas to be defined from 16 Mbytes to 2 Gbytes in size where any address translation is mapped around the normal memory management unit. Function code and read/write protection is provided. These windows are frequently used to access system memory and I/O areas without using any ATC entries or the normal memory management system.

The MMU and caches can be disabled either by external hardware using the MMUDIS* and CDIS* pins or by software and the control bits in the internal MMU and cache control registers.

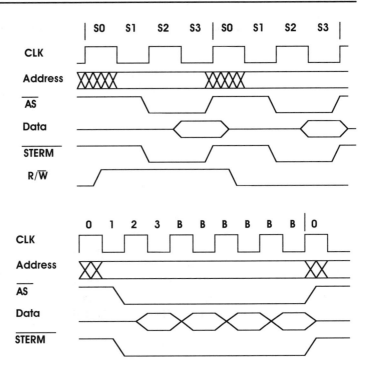

*The MC68030 synchro-
nous and burst fill bus
cycles*

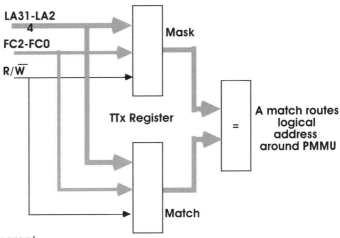

*The MC68030 transparent
windows*

The MMU supports only the PFLUSH, PMOVE, PTEST and PLOAD instructions. Attempted execution of any other PMMU instruction traps to an exception handler and an emulator if necessary. In addition, breakpoint registers, access levels, task aliasing, lockable CAM Entries and 'shared globally' entries are not supported. This may appear to be a tremendous limitation but their disappearance was due to silicon constraints and a lack of use within software. UNIX ports from the MC68020/MC68851 environment to the MC68030 were performed extremely quickly — 24 hours in some cases!

The MC68030 did not offer the performance increase seen between the MC68000 and the MC68020: it did offer about 1.4 to 1.7 times over a straight MC68020 and over 2 times over MC68020/MC68851 systems. Its biggest advantage was its integration and the use of cost effective memory to maintain the right price performance level. It is this last issue which made the appearance of a 50 MHz part feasible. Comparison of MC68020 and MC68030 designs running at 50 MHz shows that the MC68030 is more efficient due to the burst filling. Without it, there is little performance gain between a 33 MHz and a 50 MHz system, simply due to the memory delays.

While all this was going on, an alternative approach was beginning to take shape.

3   The RISC challenge

Until 1986, the expected answer to the question 'which processor offers the most performance' would be MC68020, MC68030 or even 386! Without exception, CISC processors such as these, had established the highest perceived performances. There were more esoteric processors, like the transputer, which offered large MIPS figures from parallel arrays but these were often considered only suitable for niche markets and applications. However, around this time, there started an interest in an alternative approach to microprocessor design, which seemed to offer more processing power from a simpler design using less transistors. Performance increases of over five times the then current CISC machines were suggested. These machines, such as the SUN SPARC architecture and the MIPS R2000 processor, were the first of a modern generation of processors based on a reduced instruction set, generically called reduced instruction set processors (RISC).

The 80/20 rule

Analysis of the instruction mix generated by CISC compilers is extremely revealing. Such studies for CISC mainframes and mini computers shows that about 80% of the instructions generated and executed used only 20% of an instruction set. It was an obvious conclusion that if this 20% of instructions was speeded up, the performance benefits would be far greater. Further analysis shows that these instructions tend to perform the simpler operations and use only the simpler addressing modes. Essentially, all the effort invested in processor design to provide complex instructions and thereby reduce the compiler workload was being wasted. Instead of using them, their operation was synthesised from sequences of simpler instructions.

This has another implication. If only the simpler instructions are required, the processor hardware required to implement them could be reduced in complexity. It therefore follows that it should be possible to design a more performant processor with less transistors and less cost. With a simpler instruction set, it should be possible for a processor to execute its instructions in a single clock cycle and synthesise complex operations from sequences of instructions. If the number of instructions in a sequence, and therefore the number of clocks to execute the resultant operation, was less than the cycle count of its CISC counterpart, higher performance could be achieved. With many CISC processors taking 10 or more clocks per instruction on average, there was plenty of scope for improvement.

The initial RISC research

The computer giant IBM is usually acknowledged as the first company to define a RISC architecture in the 1970s. This research was further developed by the Universities of Berkeley and Stanford to give the basic architectural models. RISC can be described as a philosophy with three basic tenets:

1.   All instructions will be executed in a single cycle.

      This is a necessary part of the performance equation. Its implementation calls for several features — the instruction op code must be of a fixed width which is equal to or smaller than the size of the external data bus, additional operands cannot be supported and the instruction decode must be simple and orthogonal to prevent delays. If the op code is larger than the data width or additional operands must be fetched, multiple memory cycles are needed, increasing the execution time.

2.   Memory will only be accessed via load and store instructions.

      This naturally follows from the above. If an instruction manipulates memory directly, multiple cycles must be performed to execute it. The instruction must be fetched and memory manipulated. With a RISC processor, the memory resident data is loaded into a register, the register manipulated and, finally, its contents written out to main memory. This sequence takes a minimum of three instructions. With register-based manipulation, large numbers of general purpose registers are needed to maintain performance.

3.   All execution units will be hardwired with no microcoding.

      Microcoding requires multiple cycles to load sequencers etc and therefore cannot be easily used to implement single cycle execution units.

Two generic RISC architectures form the basis of nearly all the current commercial processors. The main differences between them concern register sets and usage. They both have a Harvard external bus architecture consisting of separate buses for instructions and data. This allows data accesses to be performed in parallel with instruction fetches and removes any instruction/data conflict. If these two streams compete for a single bus, any data fetches stall the instruction flow and prevent the processor from achieving its single cycle objective. Executing an instruction on every clock requires an instruction on every clock.

The Berkeley model

The RISC 1 computer implemented in the late 1970s used a very large register set of 138 x 32 bit registers. These were arranged in eight overlapping windows of 24 registers each. Each window was split so that six registers could be used for parameter passing during subroutine calls. A pointer was simply changed to select the group of six registers. To perform a simple call or return simply needed a change of pointer. The large number of registers is needed to minimise the number of fetches to the outside world. With this simple window technique, procedure calls can be performed extremely quickly. This can be very beneficial for real-time applications where fast responses are necessary.

However, it is not without its disadvantages. If the procedure calls require more than six variables, one register must be used to point to an array stored in external memory. This data must be loaded prior to any processing and the register windowing loses much of its performance. If all the overlapping windows are used, the system resolves the situation by tracking the window usage so either a window or the complete register set can be saved out to external memory. This overhead may negate any advantages that windowing gave in the first place. In real-time applications, the overhead of saving 138 registers to memory greatly increases the context switch and hence the response time.

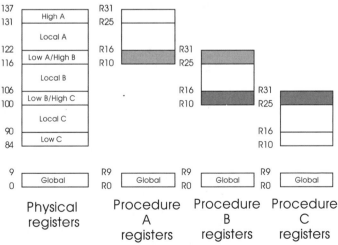

Register windowing

The Stanford model

This model uses a smaller number of registers (typically 32) and relies on software techniques to allocate register usage during procedural calls. Instruction execution order is optimised by its compilers to provide the most efficient way of performing the software task. This allows pipelined execution units to be used within the processor design which, in turn, allows more powerful instructions to be used.

However, RISC is not the magic panacea for all performance problems within computer design. Its performance is extremely dependent on very good compiler technology to provide the correct optimisations and keep track of all the registers. Many of the early M68000 family compilers could not track all the 16 data and address registers and therefore would only use two or three. Some compilers even reduced register usage to one register and effectively based everything on stacks and queues. Secondly, the greater number of instructions it needed increased code size dramatically at a time when memory was both expensive and low in density. Without the compiler technology and cheap memory, a RISC system was not very practical and the ideas were effectively put back on the shelf.

The catalysts

While CISC processors, such as the MC68020, were offering state-of-the-art performance, they were also providing examples of how far CISC instructions and addressing modes could be taken and provided ideal examples of a good principle taken to an extreme. It must be stated, however, that at the time the design incorporated many customer demands for these features and represented the current thinking. That wonderful debugging tool, hindsight, has shown that the CISC principles were possibly taken too far. The call and return module instructions, CALLM and RTM, were widely welcomed by many, but were never used in practice and, therefore, not supported on the MC68030 to allow space for other functions to be incorporated. Many of the memory indirection addressing modes can be performed faster by two separate address calculations, albeit at the expense of an extra register and instructions. As companies become more experienced in the MC68020, these anomalies were prompting an alternative approach.

Three other ingredients appeared about the same time to complete the catalytic mixture :

- Compiler technology had progressed sufficiently to be able to offer good optimisation and register tracking.

- Memory was now fast and cheap to cope with code expansion.

- The up and coming ASIC technology could be used to build the processors without the expensive investment in silicon fabrication equipment.

With all these factors coming together, it was obvious that RISC technology was now a commercial proposition and offered processors with a manyfold performance increase over their CISC counterparts. It was the culmination of all these ingredients which caused RISC technology to apparently appear overnight. This, coupled with the adoption of MIPS (million instructions per second) as the immediate performance measurement unit, led to claims which ranged from incredible to incorrect. RISC was so fashionable that almost any new processor would be described as RISC or RISC-like!

It was realised by Motorola that this new architecture offered potential and a project was launched in the mid 1980s to produce a RISC architecture. The rumours about this work started fueling a debate concerning the M68000 family. Was the MC68040 to be cancelled? Could CISC offer the performance that RISC promised? Was CISC a dead end? Was the M68000 family to be terminated? The answer to all accounts was an emphatic 'no', yet the hype surrounding RISC clouded the issue so much that these questions were continually being asked, even after three new M68000 processors had been announced!

In the end, Motorola used its CISC experience and developed a RISC architecture that combined the best of both worlds, and would be the start of a new family. To quote the 'unigram' newsletter, 'Only Motorola could take an apparently simple concept as RISC, and produce a processor that was more complex and advanced than a CISC!'. The reason for this complexity was that the M88000 architecture was designed to go beyond a single processor and looked at achieving the right balance between processor, memory and software.

The M88000 family

The MC88100 is a concurrent RISC microprocessor which achieves its performance through the use of overlapping parallel execution units. Everything about the processor is pipelined — integer and floating point calculations and data and instruction fetches. Indeed, early internal documents referred to it as a concurrent pipeline machine, however, the CPM acronym did not bring to mind a high technology image and the idea was quickly dropped!

Pipelining allows complex operations to be performed and yet give a single cycle performance. Fabricated using Motorola's HCMOS technology, the MC88100 has a Harvard architecture with separate external 32 bit address and data width program and data buses. It interfaces via a synchronous single cycle transfer pipelined protocol, PBus, capable of 132 Mbyte/second bandwidth at 33 MHz, either directly to memory systems or to the MC88200 Cache MMU chip. The MC88200 provides high speed memory caching, address translation facilities and interfaces to external memory via a synchronous M-Bus. The processor supports either Big Endian or Little Endian byte ordering. The programming model supports user/supervisor partitioning and has a 32 by 32 bit general purpose register set, with additional access to all execution unit control registers for the supervisor. For maximum performance, IEEE P754 double and single precision floating point arithmetic is supported onchip.

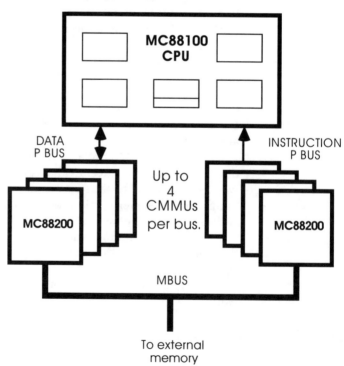

A M88000 processing node

The use of internal parallelism to overlap instruction pipelines and reduce execution times first appeared in the MC68020 and MC68030, where the bus controller, onchip caches and separate ALUs overlapped operations and reduced memory fetch, address translation times etc. Such overlapping allowed some instructions and address translations to be effectively executed in zero clock cycles. However, due to the complexity of instruction decoding, a lot of hardware is required to further develop this concept and allow the majority of instructions to benefit. The M88000, with its reduced instruction set and associated simplified decoding circuitry, is able to exploit this idea fully.

M88000 concurrent functional units

The processor unit is designed as a set of execution units running in parallel to provide performance. Each execution unit is independent and fed by its own pipeline. Three separate internal buses are used to communicate between them, enabling the triadic instruction set to be supported. On every clock edge, two source buses provide the data for the next instruction, while a third transfers the result from the preceding instruction.

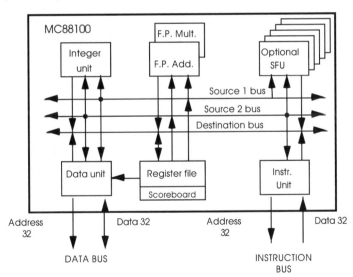

The MC88100 RISC
processor execution units

By operating concurrently, one integer instruction is executed each clock cycle and, by optimising floating point instruction sequences, these too can be overlapped to give effective single cycle execution. With the five independent execution units available, the M88000 can simultaneously access program and data memory spaces, execute an integer instruction and execute two integer or floating point arithmetic instructions. In addition, as the units are pipelined, three data and three program fetches, six floating point or four integer multiply instructions and five floating point add or subtract instructions can be in progress at any one time.

With all these parallel operations taking place, it is becoming increasingly difficult to establish a comparative MIPS rating for such processors. The number of execution units can be multiplied by their clock rate to give a burst rating. Alternatively, a time slice can be taken on a clock edge and the execution units examined. Each full stage of each execution unit is an instruction being executed and, therefore, in excess of 220 MIPS at 20 MHz could be claimed for the M88000 family. Again, there is a danger within high speed processing of becoming absorbed by MIPS without realising that they often have no reference to the quality of work performed.

Multiple execution units and optimisation

The MC88100 integer unit has its own arithmetic logic unit, bit field support and branch logic and is responsible for executing all integer, bit field and branch instructions. Integer multiply and divides are performed with help from the floating point units to provide faster performance.

The M88000 has two onchip floating point units, an adder and a multiplier which are, in themselves, fully pipelined. Although they take multiple cycles to complete, a new instruction can be dispatched on every clock edge and by using efficient compiler optimisations, single cycle floating point operations can be achieved. Both these units and the integer unit can execute concurrently. Unlike many other RISC architectures, there is no need to complete the preceding instruction before dispatching the next. Although a multiply takes about 4–6 clocks, depending on the precision, giving approximately 6 MFLOPS (Million FLOating Point calculations per Second) at 33 MHz, peak performances of 33 MFLOPS can be achieved.

Both the data and the instruction units are viewed as separate execution units with responsibility for moving data in and out of the processor. The data unit has its own address unit for address calculations and transfers data in and out of the processor via the PBUS. It can have up to four data accesses pending before the other internal execution units are stalled. The instruction unit provides instructions on a one per cycle basis and maintains three instruction pointers, XIP, NIP and

FIP, which point to the contents of the execution pipeline for efficient exception handling. It also generates the return pointer for branch and jump instructions.

Compiler optimisation for RISC processors, like the M88000, is generally based on making the most efficient use of the parallelism generated by these multiple overlapping pipelines. It identifies pipeline stalls and by re-arranging the code execution sequence, removes them while still preserving the original syntax. Such stalls can occur if execution is dependent on the availability of results from preceding instructions, or as the result of pipeline flushing during program branching. Other instructions may be inserted between dependent instructions to maintain the single cycle flow. The diagram shows the general principle involved. The effect of such optimisation can be extremely dramatic — in fast Fourier transform calculations; execution times for a loop sequence of 22 instructions, including 10 floating point operations, can be improved from over 50 clocks to just 23!

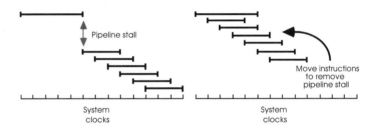

Rescheduling instructions to improve performance

With all the concurrent operations being performed, there is a potential problem when instructions are executed which require data which are not yet available from previous instruction execution. Such conditions result in an erroneous operation which would not be detected. In addition, the machine hardware integrity is totally dependent on the software environment detecting such code sequences and not executing them. This is totally unacceptable for many applications.

The difficulty facing compiler writers is in identifying these sequences from all the possible valid combinations that are available. Many optimisers are faced with a choice in approach. With machine integrity dependent on software, optimisers have to adopt algorithms which can guarantee total integrity. With hardware interlocking, an optimisation routine can be more aggressive and not worry about the one case in a hundred where the sequence integrity is destroyed.

Scoreboarding

The following diagram shows an example synchronisation problem. The first instruction performs a floating point addition where the contents of registers 1 and 2 are multiplied together and the result stored in register 3. The next instruction takes register 3, adds it to register 4 and stores the result in register 5. The problem occurs when the second instruction starts down the execution pipeline.

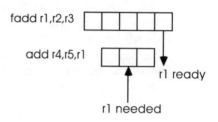

Pipeline synchronisation problem

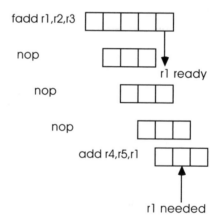

A software solution using NOPs

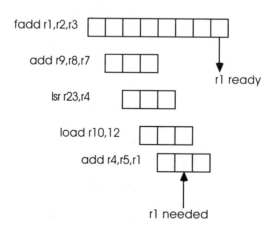

Processor corruption

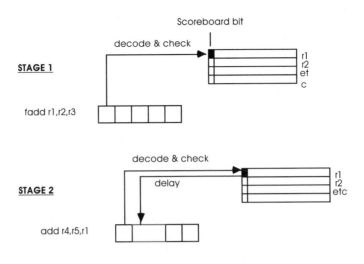

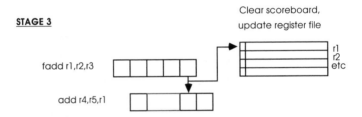

*Hardware
scoreboarding*

The value of register 3 that it uses will not be the new value and the resulting calculation will be wrong. This would corrupt the system.

There are several solutions to this. The first relies on software to recognise all such situations and insert either instructions or NOPs to delay the second instruction sufficiently.

This often wastes opportunities where instructions could have been executed and, as stated previously, makes the entire hardware integrity totally dependent on software. If the hardware changes with modifications to pipeline lengths, for instance, the software solution may no longer be valid and binary code compatibility is lost.

The MC88100 maintains synchronisation by using an internal scoreboard which is attached to the register file. Each register has a scoreboard bit which is set as a result of decoding an instruction which modifies the register contents. A pipeline which uses that information is delayed clock by clock until it is available and is fed directly from the pipeline before updating the register file. This reduces any delays to a minimum. As a result, compiler optimisers can concentrate on optimising and not maintaining system synchronicity.

Delayed branching

With any pipelined architecture, there is always associated stalling due to program branching or jumps. This is caused by an unwanted instruction being prefetched into the pipeline, which subsequently needs flushing to remove it. This obviously reduces impact. One solution is to have a branch prediction cache which stores the two instructions immediately after the branch. This cache mechanism is only effective on software loops and effectively stores the instructions that would be taken if the flow changed. When this happens, the instructions are already available and can be fed into the execution unit. This mechanism takes large areas of silicon to implement and its performance is heavily dependent on software, cache size and hit rates.

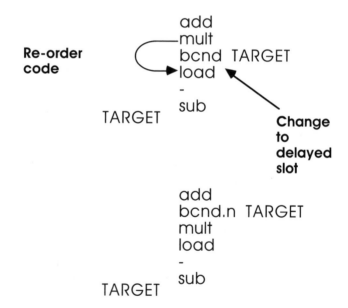

Delayed branching

Delayed branching is effectively a software optimising technique which is applicable to all software, irrespective of a looping structure or not. It consumes little silicon real estate and can provide good performance increases. The technique works by reordering the instruction sequence. The preceding instruction is moved to immediately after a branch op code and executed while the next instruction is fetched.

The unwanted instruction that would normally be inserted into the pipeline can now be executed and the pipeline not flushed to remove it. Using this technique, it is common to achieve 20 MIPS at 20 MHz.

The MC88100 programming model

The MC88100 processor supports the concept of separate supervisor and user spaces. The user programming model consists of a register file with 32 x 32 bit wide general purpose registers, r0 to r31. These can perform data and addressing operations without any restriction. Register r0 is read only and is hardwired to contain the value zero to provide a zero reference for the instruction set. In addition, there are two control registers associated with the floating point execution unit. A major difference is the lack of a condition code register containing flags. This appears to be an omission within the architecture on first sight. In reality, the results of condition code testing through a compare operation are stored in one of the other registers.

The supervisor has access to the the user model and to the control registers for the integer and floating point execution units. These registers provide the processor status and exception information.

r0	ZERO
r1	SUBROUTINE RETURN POINTER
r2	CALLED PROCEDURE PARAMETER REGISTERS
r9	
r10	CALLED PROCEDURE TEMPORARY REGISTERS
r13	
r14	CALLING PROCEDURE RESERVED REGISTERS
r25	
r26	LINKER
r27	LINKER
r28	LINKER
r29	LINKER
r30	FRAME POINTER
r31	STACK POINTER

The M88000 user programming model

cr0	PID	PROCESSOR IDENTIFICATION REGISTER
cr1	PSR	PROCESSOR STATUS REGISTER
cr2	TPSR	TRAP PROCESSOR STATUS REGISTER
cr3	SSBR	SHADOW SCOREBOARD REGISTER
cr4	SXIP	SHADOW EXECUTE INSTRUCTION POINTER
cr5	SNIP	SHADOW NEXT INSTRUCTION POINTER
cr6	SFIP	SHADOW FETCH INSTRUCTION POINTER
cr7	VBR	VECTOR BASE REGISTER
cr8	DMT2	TRANSACTION REGISTER #2
cr9	DMD2	DATA REGISTER #2
cr10	DMA2	ADDRESS REGISTER #2
cr11	DMT1	TRANSACTION REGISTER # 1
cr12	DMD1	DATA REGISTER # 1
cr13	DMA1	ADDRESS REGISTER #1
cr14	DMT0	TRANSACTION REGISTER # 0
cr15	DMD0	DATA REGISTER #0
cr16	DMA0	ADDRESS REGISTER #0
cr17	SR0	SUPERVISOR STORAGE REGISTER # 0
cr18	SR1	SUPERVISOR STORAGE REGISTER # 1
cr19	SR2	SUPERVISOR STORAGE REGISTER # 2
cr20	SR3	SUPERVISOR STORAGE REGISTER # 3

*The M88000 integer
unit control registers*

Handling exceptions

The processor has a similar USER/SUPERVISOR concept to that of the M68000 family. Normally the processor
executes code in the USER state and only moves into the
SUPERVISOR state when an exception or some error condition is encountered (e.g. an interrupt or bus error). The biggest
difference between the M88000 and M68000 families is the
way the processor context or internal state is preserved. With
the M68000 family, this is done by building an external stack
frame which can take up to approximately 40 separate memory
transfers. The M88000 family constantly updates a set of
shadow registers so that the context is immediately preserved
when an exception happens. The updating is simply stopped
and this removes the delay.

The architecture of the M88000 supports two types of
exception — precise and imprecise. Precise exceptions are
those that require an instruction to restart and therefore the
full information concerning the exception is needed. Imprecise exceptions are those such as divide by zero, where the

instruction will be discontinued and the program flow started elsewhere. A range of exceptions similar to that of the M68000 family is supported via a relocatable vector table. Each vector references a 64 bit wide location which contains two instructions. These are usually a jump to the handler followed by the first handler instruction in a delayed branch mode. This allows the jump to be completed without causing a pipeline stall.

The MC88100 has a single interrupt pin which is used to signal an external interrupt. If this level sensitive pin is asserted for two successive falling clock edges, an interrupt is internally signalled and the processor context saved to an internal set of shadow registers — not to external memory. The current instruction does not have to complete and the only variable in the latency calculation is the time taken to complete any data and instruction unit memory accesses. Worst case is four operations for the data unit and two for the instruction unit. Typical values are six clock cycles which, at 25 MHz, is less than 0.25 microseconds. This is at least an order of magnitude faster than more conventional architectures.

The MC88100 instruction set

The instruction set is based on a load/store memory operation RISC architecture with triadic operations providing non-destructive data usage. By specifying two sources and a different destination, the source data is maintained and available for future use. This reduces the software overhead a diadic instruction set needs by preserving data where the destination is one of the two sources. All instructions are 32 bits wide and aligned. The set has 51 instructions that support integer and floating point arithmetic, logical and flow control, load/store instructions and bit field instructions. The latter are normally associated with a complex instruction set—their inclusion improves software performance without comprising the processor design goal of hardwired logic and therefore offers a clear advantage.

The M88000 instruction set has been optimised rather than reduced. This means that many conventional CISC instructions can be replaced on a one-to-one basis, which helps to keep code expansion ratios as close to this ratio as possible. With CISC processors fast approaching single cycle execution, RISC instruction sets cannot be so reduced that they take more instructions than CISC instructions take clocks to execute. If this happens, a RISC processor may offer an outstanding MIPS figure yet actually perform less work than expected! The importance of the instruction set is discussed in a later chapter.

Addressing data

Unlike many RISC architectures, which have only one address mode and require addresses to be calculated prior to every external access, the M88000 instruction set supports seven.

Addressing Mode		Syntax
Register Indirect with Unsigned Imediate		rd,rs1,imed16
Register Indirect with Index		rd,rs1,rs2
Register Indirect with Scaled Index		rd,rs1 [,rs2]
Register with 9-bit Vector Number		m5,rs1,vec9
Register with 16-bit Signed Displacement		m5,rs1,d16
		b5,rs1,d16
Program Counter Relative	(26-bit Signed Displacement)	d26
Register Direct		rs2

The M88000 addressing modes

Probably the most useful is the register indirect with an index, which can be scaled. This allows a pointer to be assigned to the head of a table or set of tables and allows other registers to index into it, dramatically reducing the number of address calculations and instructions required. Simply changing the main pointer register to that of another table allows rapid searching of tables and linked lists. This reduces the number of instructions needed per function and increases throughput.

Fetching data

Six data types are supported by the M88000 instruction set as shown. Bit field data can be from 1 to 32 bits in size. Integers can be either byte, halfword or word in length — yes, the M88000 defines 32 bits as a word while the M68000 family defines 16 bits as a word and 32 as a long word! (The Motorola DSP56000 family has a 24 bit word, just to add to the confusion.) Floating point values are either 32 or 64 bits in size.

The only exception to the load-store architecture used by the M88000 is the xmem instruction which exchanges a memory contents with a memory value in two successive indivisible cycles. This provides basic multiprocessor support. The load-store model effectively removes the memory access time from the instruction execution time. Such operations take only three instructions and, therefore, three cycles to complete. Direct memory modifying architectures take three instruction times plus the time for two memory accesses.

Data Type	Represented As
Bit Fields	Signed and Unsigned Bit Fields (1 to 32 bits)
Integer	Signed and Unsigned Byte (8-bits)
	Signed and Unsigned Half-word (16-bits)
	Signed and Unsigned Word (32-bits)
Floating Point	IEEE P754 Single Precision (32-bits)
	IEEE P754 Double Precision (64-bits)

The M88000 data types

Further consideration must be given to how much data can be fetched and how this fetching is performed. Most compiler-generated data, like 'C' variables such as char, short and int, are 8 or 16 bits in length, yet current RISC architectures are 32 bit in data size. While many RISC architectures can deal with the corresponding byte and halfword data sizes internally, they have to fetch such data using a multiple instruction sequence. The first instruction calculates the address and loads it into the address register, a 32 bit wide word of data is fetched into a register and, finally, a third instruction is needed to extract the relevant byte or halfword from that register and store it in yet another register. The MC88100 has four data byte enable signals which allow it to fetch individual bytes and halfwords directly in a single cycle. This again reduces code expansion, the number of cycles needed to perform the task and, ultimately, improves the system performance.

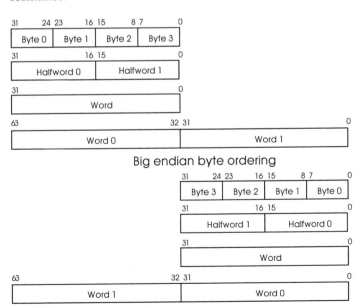

Big endian byte ordering

Little endian byte ordering

Byte ordering

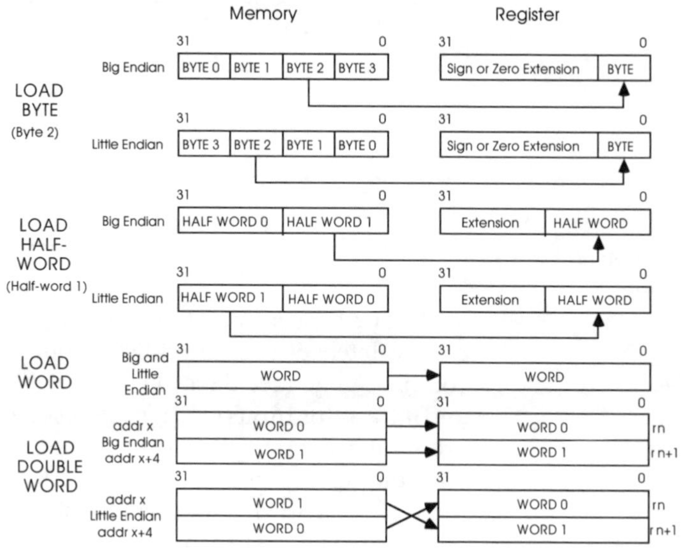

	Memory		Register	

The diagram shows Load Byte, Load Half-word, Load Word, Load Double Word transfers between Memory and Register for Big Endian and Little Endian.

LOAD BYTE (Byte 2)

Big Endian — Memory: BYTE 0 | BYTE 1 | BYTE 2 | BYTE 3 → Register: Sign or Zero Extension | BYTE

Little Endian — Memory: BYTE 3 | BYTE 2 | BYTE 1 | BYTE 0 → Register: Sign or Zero Extension | BYTE

LOAD HALF-WORD (Half-word 1)

Big Endian — Memory: HALF WORD 0 | HALF WORD 1 → Register: Extension | HALF WORD

Little Endian — Memory: HALF WORD 1 | HALF WORD 0 → Register: Extension | HALF WORD

LOAD WORD

Big and Little Endian — Memory: WORD → Register: WORD

LOAD DOUBLE WORD

addr x / addr x+4 Big Endian — Memory: WORD 0 / WORD 1 → Register: WORD 0 (rn) / WORD 1 (rn+1)

addr x / addr x+4 Little Endian — Memory: WORD 1 / WORD 0 → Register: WORD 0 (rn) / WORD 1 (rn+1)

Data transfers

The diagram shows how data is transferred from memory to the internal registers. The processor can be set up on reset to use either big or little endian byte ordering. These two schemes differ in which is the most significant byte or 16 bit halfword within a 32 bit word. With big endian ordering, the most significant byte is located at bits 24–31. With little endian the same byte is located at bits 0–7. This in itself is a minor difference, although the implications for a system are immense. If an instruction tests bit 18, different data would be accessed giving differing results within a system that appeared to be compatible. Often C programs which try to be portable across many systems fail to meet this goal because of assumptions about byte ordering. The byte ordering within the MC88100 processor registers is not changed.

MC88100 external functions

The functional pinout of the device is shown. The bulk of the signals are concerned with the two PBUS memory interfaces which form the Harvard architecture. The processor clock is supplied via the CLK pin and the internal phase lock looping is enabled via the PLLEN signal. This reduces the problem associated with clock skewing between multiple processors or cache memory management units. There is a single reset line and a single interrupt line. The PCE signal allows the processor to be configured either as a master or a checker within a fault tolerant pair. (Fault tolerance is discussed further in Chapter 9.)

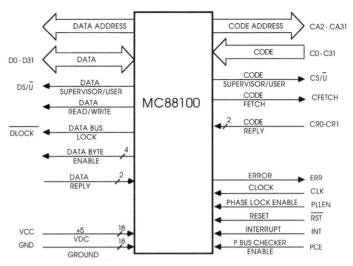

The MC88100 signal pin-out

The PBUS signals are essentially the same for both the data and code buses and use the same protocol. The instruction bus, however, cannot write to its memory space and all accesses are 32 bit wide, hence the missing CA0 and CA1 signals. The data side can perform read/write operations, indivisible cycles and with the byte enable strobes access 8, 16 and 32 bit data. Both buses support the USER/SUPERVISOR partitioning. The two pairs of control registers, DR0-1 and CR0-1, tell the processor whether to insert wait states and whether the transfer has completed successfully or incorrectly.

DATA P BUS		INSTRUCTION P BUS	
Signal	**Description**	**Signal**	**Description**
DA2 - DA31	Data Address Bus	CA2 - CA31	Code Address Bus
D0 - D31	Data Bus	C0 - C31	Code Bus
DS/$\overline{\text{U}}$	Data Supervisor/User Select	CS/$\overline{\text{U}}$	Code Supervisor/User Select
DR/$\overline{\text{W}}$	Data Read/Write - asserted for read, negated for write		(no corresponding signal)
$\overline{\text{DLOCK}}$	Data Bus Lock - asserted for cache lock		(no corresponding signal)
DE0 - DE3	Data Byte Enables	CFETCH	Code Fetch - asserted for fetch
DR0 - DR1	Data Reply 0 and 1 - DR0 DR1 0 0 Reserved 0 1 Success 1 0 Wait 1 1 Fault	CR0 - CR1	Code Reply 0 and 1 - CR0 CR1 0 0 Reserved 0 1 Success 1 0 Wait 1 1 Fault

The PBUS signals

Single-cycle memory buses

The fine grain parallelism used in the MC88100 requires external memory systems to provide data on every clock edge in order to prevent pipeline stalls reducing performance. This effectively means that at 20 MHz memory is accessed every 50 ns and at the higher clock speeds, such as 40 MHz, this increases to every 25 ns. This places tremendous design constraints on the external memory systems. To maintain the constant flow of instructions into the processor requires a separate bus with a very high bus utilisation (95–100%), hence the Harvard bus architecture with separate data and instruction buses. The insertion of a single wait state halves system performance, two wait states reduce it to a third and so on. A 20 MIPS machine can be reduced to 6 MIPS, which, considering the poorer instruction set and addressing modes available to most RISC architectures, may actually performs less work than a well designed CISC alternative. RISC systems may have simpler processors but their memory systems are more sophisticated, often requiring cache memory subsystems.

The processor memory interface, PBUS, uses a pipelined protocol to maintain a single cycle instruction flow using an effective three cycle access.

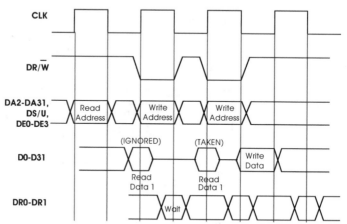

The PBUS protocol during a READ-WAIT-WRITE cycle

The bus cycle starts with an address with the USER/ SUPERVISOR status, byte strobes and read/write status. In the example shown, a read takes place. On the next clock, the data is read in from the data bus by the processor. This always occurs due to the pipelining, however, the data is not used until the reply signals are received on the third clock. During the second clock, the address information for the next cycle is driven on to the bus. The reply to insert wait states is passed back on the third clock. The address data is maintained on the bus and the data bus is sampled again on the next clock. This time, the data is good and a success is indicated on the subsequent clock. The next operation is a write, the data appears and again success is indicated by the control signals.

Each access actually consists of three stages but the pipelining allows each stage to be completed on every clock cycle, resulting in an effective single-cycle performance. Multiple cycles simply overlap each other. Any memory design must therefore latch the address. The PBUS is normally used with direct connections to MC88200 CMMUs but can be connected directly to memory if required.

The biggest restriction in such direct memory designs is upgrading the system clock speed. Most of today's M68000 family designs are capable of running different speed processors on the same hardware by simply replacing the processor and keeping the memory system and peripherals the same. The insertion of wait states does not degrade the system performance too much. With a RISC architecture, both the processor and its cache memory need upgrading to prevent wait states on the processor buses from destroying performance. Faster processors need faster caches which need faster memory chips. The problem comes with the interface logic and buffering that is also needed. The buffering delays of a few nanoseconds become very significant as the cycle times become shorter. In addition, the effect of impedance mismatch and capacitive loading due to design layout and printed circuit board track lengths start exaggerating the problems. In effect, it will be virtually impossible to speed upgrade such designs without redesigning the board. The only sensible method of solving these problems is to integrate the various functions on to one piece of VLSI manufactured in the same technology. With both the processor and its cache memory system using similar fabrication techniques, they can easily track each other as the faster versions become available. A board upgrade simply becomes a processor and CMMU replacement without a redesign.

MC88200 cache MMU

The MC88200 Cache/Memory Management Unit provides 16 kbytes of one cycle access, four ways set associative cache memory together with full support for segmented and demand paged memory management schemes. By integrating memory caches with memory management functions, translation delays can be removed by overlapping cache accesses with address translations and thus the use of memory management does not reduce system performance. This idea is present on the MC68030 but again has been further developed within the M88000 family.

Memory management functions

Integrating memory management and cache memory on a single chip allows their operations to be overlapped so there is no penalty for their use. This is done by performing the cache lookup and address translation in parallel. The logical address bus is split, allowing the lower bits to be used within the cache lookup mechanism while the address translation is done. When the correct physical address is ready, it makes the final data selection from the potential cache entries that the lookup has generated. If data and the address translation descriptor are cached on chip, the data is supplied with no wait states to the processor. If the the descriptor is missing, the CMMU performs an external table walk into external memory to fetch it. If the data is not present, it is fetched from memory.

The CMMU supports various memory granularity:

- 4 Gbytes user space
- 4 Gbytes supervisor space
- Segment (4 Mbytes on 4 Mbyte boundary)
- Block (512 Kbytes on 512 Kbyte boundary)
- Page (4 Kbyte on 4 Kbyte boundary)

Granularity is controlled by three descriptors: page , segment and block. In addition to the more normal read/ write user/supervisor protection bits, various cache coherency mechanisms, such as cache inhibit, write back, etc. can be dynamically invoked. By coupling the cache control with the memory management in this way, it is very easy to configure the system for optimum performance.

In addition to the cache memory, there are two caches for high speed address translation:

- the page address translation cache (PATC), containing 56 descriptors, and

- a block address translation cache (BATC), with a 10 entry table of memory block descriptors.

In most systems, the operating system uses the BATC, while applications use the PATC. This prevents operating system descriptors from being overwritten by application entries as tasks are loaded and run.

Area

Sup'v. Segment Table Base	WT	0	G	a	0	0	U	WP	V
User Segment Table Base	WT	0	G	a	0	0	U	WP	V

Segment

Page Table Base	WT	SP	G	a	0	0	U	WP	V

Page

Page Frame Base	WT	SP	G	a	0	M	U	WP	V

V	Valid	a	Cache Inhibit
WP	Write Protect Enable	G	Global - Snoop
U	Used	SP	Supervisor Only
M	Modified	WT	Write Through

MC88200 MMU
descriptors

Cache coherency

The 16 Kbyte memory cache is on the physical memory side, with the advantage of no flushing on every context switch and, therefore, maintains high hit rates of about 97–99%, although such figures are highly software dependent. A typical processing node may use up to eight MC88200s, split between the data and program buses.

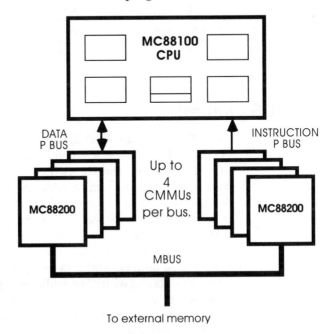

An MC88100/88200
processor node

With a maximum number, a M88000 processor node would have 128 Kbytes of wait state free cache along with 448 PATC and 80 BATC translation entries. This provides an elegant solution to the von Neumann bottleneck, which is often a limiting factor in both RISC and CISC architectures. By comparison, the MC68030 has 512 bytes of cache and a 22-entry translation cache.

The MC88200 provides a hardware solution for the problem of stale data cache coherency. Entries can be allocated either with a copy back policy to reduce external traffic or written through. There is a bus snoop protocol which monitors the external bus and updates the cache when a data entry is modified externally. This function is again essential to any high performance system which uses either multiple processors or DMA to prevent corruption and the resulting system crashes. Caches, and their coherency are described in greater detail in a later chapter.

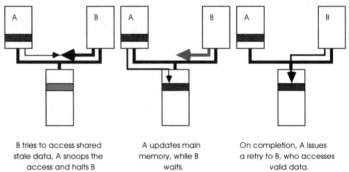

B tries to access shared
stale data, A snoops the
access and halts B

A updates main
memory, while B
waits.

On completion, A issues
a retry to B, who accesses
valid data.

Cache coherency
versus bus snooping

The MBUS protocol

Each MC88200 CMMU communicates to main memory via the MBUS. This is a synchronous pipelined bus which supports both single and block transfers and concurrent arbitration. The bus has an arbitration mechanism which allows designers to use a single bus structure or maintain the Harvard architecture of the processor. At 20 MHz, it is capable of transfers up to 64 Mbytes/second.

Due primarily to a pin-out restriction, the bus uses a multiplexed address and data bus. The control and status signals, associated with it have different meanings depending on which phase of the cycle is current. During the first 'request' phase, indicated by C0 = 1, the address is generated along with other information describing the bus access. This could be a non-cached global access involving a locked sequence, for example. The memory or subsystem responds either with a success, wait or error via the system status signals, SS0-3*. Wait states can be inserted to allow extra address decoding time.

Once completed, the second 'data' phase commences. Here, some control signals change meaning and become byte enable strobes to indicate whether this is the last transfer in the sequence. The data appears on the bus and the system status signals can show success or error or insert wait states. There is a special signal, however, which indicates that the memory cannot accept any more transfers and that this should be the last. The combination of these SS0* and C1 signals allow block transfers to be controlled by either the CMMU or the memory. Parity generation and checking can be selected for the bus and its control signals.

CONTROL SIGNAL	REQUEST PHASE	DATA PHASE
C0	REQUEST = 1	REQUEST = 0
C1	Intent to Modify	End of Request (EOR)
C2	Read/$\overline{\text{Write}}$	Read/$\overline{\text{Write}}$
C3	Locked	Byte Strobe 3 (I31-I24)
C4	Cache Inhibit	Byte Strobe 2 (I23-I16)
C5	Global	Byte Strobe 1 (I15-I8)
C6	reserved	Byte Strobe 0 (I7-I0)

SYSTEM STATUS SIGNAL	REQUEST PHASE	DATA PHASE
$\overline{\text{SS3}}$	ERROR	ERROR
$\overline{\text{SS2}}$	RETRY	RETRY
$\overline{\text{SS1}}$	WAIT	WAIT
$\overline{\text{SS0}}$	reserved	End of Data (EOD)

The MBUS control and status signals

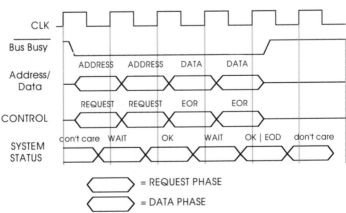

An MBUS cycle with wait states

M88000 master/checker fault tolerance

One feature that has not been previously seen within the MC68000 family, is the ability to run a second processor node in a fault checking shadow mode without any additional comparator and/or synchronising logic. Both the MC88100 and MC88200 have an external pin which, when asserted, forces the processor to execute internally without driving the external buses. Instead, the external signals from the master processor are compared by the checker processor and, if there is a discrepancy, an error signal is asserted. The checking circuits can also detect external bus faults on power up. Further protection is given by the use of four parity bits on the MC88200 MBus interface to external memory. This fault detection scheme has no impact on performance whatsoever and allows redundant MC88100 nodes to be easily designed.

Future enhancements

With only the MC88100 and MC88200 parts initially available, it may appear to be a little difficult to see how the family is likely to progress. However, if the architecture is examined in a little more detail some ideas do come to mind.

The MC88100 diagram shows a group of special function units (SFUs). The floating point unit is SFU 1 and up to six more can be added at a later stage. All have their own internal interface to the register file and scoreboard and 256 unique op codes are predefined for them.

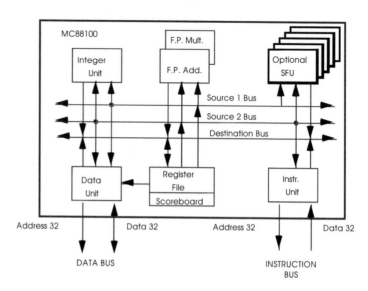

Special function units

If such an op code is executed on a MC88100, it causes an exception and allows software to emulate the function. This situation is extremely similar to the F-line and A-line exceptions on the MC68000. They allowed extra coprocessors to be added on future generations.

If extra execution units are to be added, the next obvious step from single cycle execution is to start executing multiple instructions per clock. For this to perform correctly, a form of hardware interlock is needed to determine when instructions can be executed without corrupting the software precedence. This can be caused by an instruction wanting the result of a preceding instruction or by multiple instructions all competing for the same resource like a condition code or status register — but the MC88100 already has a scoreboard mechanism and has no condition code registers or status registers. Is this just coincidence? All that was certain was that the next generations were already under development and that they would execute multiple instructions per clock. However, as we will see in the next chapter, this requires additional architectural features and tricks.

4   RISC wars

RISC versus CISC

The late 1980s and early 1990s were a very difficult time for both users and suppliers of high performance RISC and CISC processors. Although RISC had promised to offer faster, cheaper and less power hungry processors, the established Motorola M68000 and Intel 80x86 families seemed to be able to compete and provide the additional benefit of a vast software base that was already established. It was very difficult for anyone to predict which architecture would dominate and more importantly, when a change from CISC to RISC should be made. The problem was made even worse by the road maps shown for both the Motorola and Intel CISC architectures which showed a never ending supply of MIPS.

The main problem for all the RISC purveyors at this time was the fact that no single large computer vendor had endorsed the technology. It was true that many workstation manufacturers had embraced the technology — or several in some cases — but this was not really a sufficient market to support the number of suppliers and the number of different RISC offerings that were present.

The market statistics show the problem. RISC processors were sold in the tens and hundreds of thousands compared to the millions of M68000 and 8086 processors sold every year. The total sales of all RISC processors during this time period would only equal one month's CISC production. It was obvious that unless a large computer manufacturer used RISC technology — one that had a major presence in the PC market — RISC processors would have to rely on the embedded controller market to provide the economies of scale. The problem faced in this market was again one of providing migration of the existing CISC software base and competetion from the new integrated procesors like the Motorola MC68300 series.

This was also the period of coming and going of alliances and consortia as the various manufacturers and users jostled for position.

There was one point that was becoming clear: CISC processors like the MC68040 were offering single cycle performance and similar performance levels compared to the first generation RISC processors such as the MC88100. As a result, there was a lot of activity to bring out the next generations of RISC to maintain the performance leadership and thus their competitiveness.

Enter the MC68040

The MC68040 was the first of the MC68000 CISC processor family to achieve single cycle instruction execution. Fabricated in 0.8 micron HCMOS technology and using 1.2 million transistor sites, it incorporates separate integer and floating point units, giving sustained performances of 20 integer MIPS and 3.5 double precision Linpack MFLOPS, dual 4 Kbyte instruction and data caches, dual memory management units and an extremely sophisticated bus interface unit.

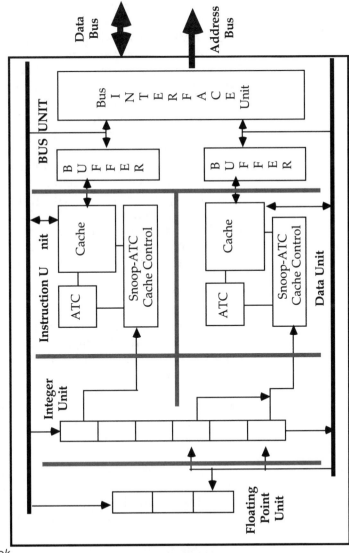

The MC68040 block diagram

The block diagram shows how the processor is partitioned into several separate functional units which can all execute concurrently. It features a full Harvard architecture internally and is remarkably similar, at the block level, to the M88000 RISC processor.

Its design was revolutionary rather than evolutionary. It takes the ideas of overlapping instruction execution and pipelining to a new level for CISC processors. The floating point and integer execution units work in parallel with the onchip caches and memory management to increase overlapping so that many instructions are executed in a single cycle — giving it its performance. Examination of the pin-out reveals a large number of new signals. One major difference with the MC68040 is its high drive capability. The processor can be configured on reset to provide drive currents of either 5 or 55 ma per bus or control pin. This removes the need for externals buffers, reducing chip count and the associated propagation delays which often afflict a high speed design.

The 32 bit address and 32 bit data buses are similar to its predecessors, although the signals can be optionally tied together to form a single 32 bit multiplexed data/address bus.

The user programmable attributes, UPA0 and UPA1, are driven according to the setting of two bits within each page descriptor. They are used by the onboard memory management units to enable the MC68040 Bus Snooping protocols and also to give additional address bits and provide software control for external caches and other such functions.

The two size pins, SIZ0 and SIZ1, no longer indicate the number of remaining bytes left to be transferred as they did on the MC68020 and MC68030, but are used to generate byte enables for memory ports. They now indicate the size of the current transfer. Dynamic bus sizing is supported via external hardware, if required. Misaligned accesses are supported by splitting a transfer into a series of aligned accesses of differing sizes. The transfer type signals, TT1 and TT2, indicate the type of transfer taking place and the three transfer modifier pins, TM0–TM2, provide further information. These five pins together effectively replace the three function code pins. The TLN0 and TLN1 pins indicate the current longword number within a burst fill access.

The synchronous bus is controlled by the master and slave transfer control signals. Transfer start (TS*) indicates a valid address on the bus, while the transfer in progress (TIP*) signal is asserted during all external bus cycles and can be used to power up/down external memory to conserve power in portable applications. These two master signals are complemented by the slave signals — transfer acknowledge (TA*) successfully terminates the bus cycle, while transfer error acknowledge (TEA*) terminates the cycle and the burst fill as a result of an error. If both these signals are asserted on the first access of the burst, the cycle is terminated and immediately re-run. On the 2nd, 3rd and 4th accesses, a retry attempt is not allowed and the processor simply assumes that an error has occurred and terminates the burst as normal.

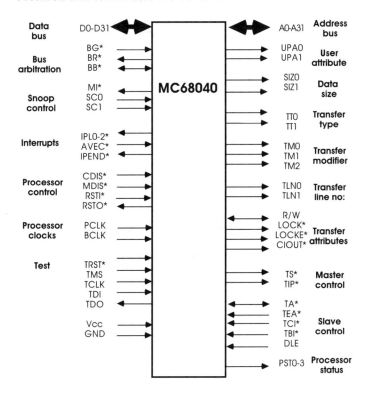

The MC68040 pin-out

The processor can be configured to use a different signal, data latch enable (DLE) to latch read data instead of the rising edge of the BCLK clock. The internal caches and memory management units can be disabled via the CDIS* and MDIS* pins, respectively.

The four processor status signals, PST0–PST3, provide information on exactly what the processor is doing internally. This data can identify supervisor and user activities and usually bears little relationship to the current bus cycle. This is provided for instrumentation support.

JTAG testing is supported via the TEST signals. This allows bit patterns to be output from the device under controlled conditions to test peripherals etc. without having to isolate the processor.

PST3-0	Internal status
0000	User start/continue current instruction
0001	User end current instruction
0010	User branch taken and end current instruction
0011	User branch not taken and end current instruction
0100	User table search
0101	Halted state (double bus fault)
0110	Reserved
0111	Reserved
1000	Supervisor start/continue current instruction
1001	Supervisor end current instruction
1010	Supervisor branch taken and end current instruction
1011	Supervisor branch not taken and end current instruction
1100	Supervisor table search
1101	Stopped state (supervisor instruction)
1110	RTS executed
1111	Exception stacking

The PST signal meanings

There are two clock signals, PCLK and BCLK. The processor clock, PCLK, runs at twice the speed of the bus clock, BCLK, allowing the internal nodes and execution units to run at twice the speed. The 25 MHz MC68040 actually executes internally at 50 MHz.

The instruction execution unit

The instruction execution unit consists of a six-stage pipeline which sequentially fetches an instruction, decodes it, calculates the effective address, fetches an address operand, executes the instruction and finally writes back the results.

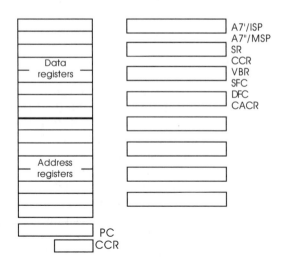

The MC68040 integer unit programming model

To prevent pipeline stalling, an internal Harvard architecture is used to allow simultaneous instruction and operand fetches. It has been optimised for many instructions and addressing modes, so that single cycle execution can be achieved. The early pipeline stages are effectively duplicated to allow both paths of a branch instruction to be processed until the path decision is taken. This removes pipeline stalls and the subsequent performance degradation. While integer instructions are being executed, the floating point unit is free to execute floating point instructions.

The programming model is essentially the same as that of the MC68030, although only two bits are used in the CACR register to enable or disable either of the two onchip caches.

The floating point execution unit

The MC68040 has its own floating point unit on chip offering >3.5 Mflops performance. It provides hardware support for the four arithmetic functions and executes the other transcendental functions via a software emulation package.

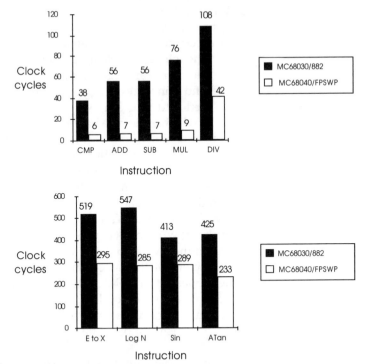

MC68040 FPU instruc-
tion timings

This combination provides support for the existing MC68881/882 floating point coprocessor instruction sets. The performance improvements are shown.

Memory management

The MC68040 has two memory management units, one allocated for data and the other for instructions. The MMUs are a subset of the MC68030 MMU and are extremely simple to program and control. Two root pointers are supported, one each for user and supervisor use. Page sizes of either 4 or 8 Kbytes are selected by bits within the translation control register and are supported via a fixed three-level set of table descriptors. To prevent external table walking, each MMU has a 64-entry 4-way set associative address translation cache which contains the most recently accessed page descriptors. The MC68030 concept of transparent windows has been utilised, with four transparent windows being shared between the two MMUs to allow memory areas to be mapped without using the ATCs.

Instruction and data caches

To keep instructions and data flowing into the execution units, the MC68040 has two four-way set associative physical caches, each 4 Kbytes in size. One cache is dedicated to instructions, the other to data. The caches provide the single cycle access necessary to maintain the single cycle execution within the integer and floating point unit pipelines.

The caches are organised as four sets, each with 64 lines per set and a line size of 16 bytes. Each line has a valid bit and is divided into four longwords, each with a 'dirty' bit to indicate whether the data has been modified. The cache is on the physical side of the onchip memory management units and thus does not need flushing on a context switch. This is important to maintain performance. Each cache is coupled with its associated MMU so that ATC lookups are performed in parallel with cache lookups and there is no performance delay. If there is a data miss and a line must be replaced, any dirty data is written out to main memory and the replacement line brought in as a burst fill of a whole cache line. This is a development of the synchronous burst memory interface, which appeared on the MC68030 and uses page mode memory to access memory on cycles 2, 3 and 4 on successive cycles, although the first access may incur wait states.

The caches support four main coherency schemes:

- cacheable with data writethrough;
- cacheable with copy back;
- non-cacheable; and
- special access.

Selecting 'cacheable with writethrough' via a page descriptor forces all data writes through to main memory, thus enforcing data coherency, i.e. cache and memory data are identical. This couples the processor unnecessarily to slow memory and so the copyback scheme is also supported. Here, cache data is updated but the main memory is not immediately written to. This reduces the number of external memory accesses and associated delays. Cache data is marked as dirty, using a tag bit, and is only written out as the result of a cache flush or if it is to be replaced. To ensure data coherency, bus snooping is supported which monitors the external bus to detect any access to an address where the main memory data is stale.

Non-cacheable is straightforward and is used to prevent data being cache. This would be needed when accessing I/O, for example. Special accesses, such as table walks, exception stack frame building and the new MOVE16 instruction which moves 16 bytes anywhere within the memory space, are not cached.

The bus interface unit

The key to maintaining the internal execution units is to provide extremely fast accesses to memory so that the internal caches can be updated quickly and any delays through to main memory are kept as short as possible. The basic interface uses a 32 bit data and address bus which can be either multiplexed or non-multiplexed as required. The interface is synchronous but allows the inclusion of wait states, enabling it to communicate to different speed peripherals or memory. It uses a burst fill interface, similar to that of the MC68030, where the first access takes a minimum of two cycles and the remaining three are single cycle. The processor always assumes that the external ports are 32 bit, aligned and capable of supporting burst filling. Where this is not possible, as with a peripheral, burst filling can be inhibited by asserting the TBI* signal during the first access. Similarly, asserting the TCI* signal inhibits internal data caching.

The BIU is intelligent enough not to discard valid cache lines until the replacement line has been completely fetched from memory. This is achieved by flushing the line into a buffer, performing the external access and then moving the cache line into the cache RAM. If an error occurs during the external fetch, the old line is reinstated into the cache. If the error occurs while fetching data needed by the execution units, a full bus error exception is taken. If the error happens during the remainder of the burst fill and the data is not needed, the good data from the earlier cycles is taken and the bus error exception deferred. In practice, data is not cached, ensuring that it is always fetched again from external memory. If the error then occurs, the exception is taken. This effectively provides an additional automatic retry mechanism to the available hardware retry. In addition, the BIU does not use burst fills for accesses that do not allocate data cache on a read miss, such as table searches, vector fetches, exception unstacking and writing a single dirty long word from a cache line. In these cases, it performs a single byte, word, or longword transfer.

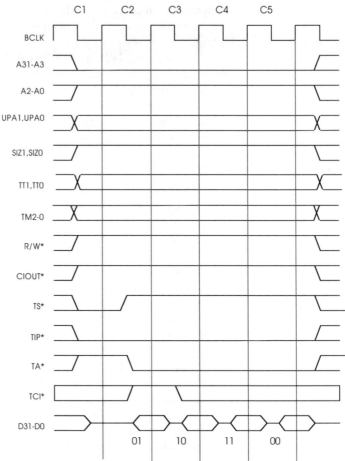

A typical MC68040 burst fill access

The bus cycle commences by the MC68040 issuing a bus request, BR*, waiting for bus grant, BG*, to be asserted and bus busy, BB*, to be negated before claiming ownership of the bus. Unlike its predecessors, the MC68040 does not automatically have the highest priority within any system. The priority and arbitration scheme is user defined and can have any implementation.

Associated with the arbitration mechanism is the locked cycle support for indivisible cycles. The LOCK* signal indicates an indivisible bus cycle to the arbiter so it can enforce the locked sequence. The LOCKE* signal is issued on the last transfer of any locked sequence, allowing bus arbitration to overlap the last transfer and indicating when to remove the BG* signal without breaking a locked transfer. The processor supports implicit bus ownership where the arbiter asserts BG* to the processor but neither it nor another processor needs it. In this case, the processor implicitly owns the bus and drives the address bus and transfer attributes with undefined values.

Exception handling

The MC68040 uses the standard M68000 seven-level external interrupt mechanism with both autovectored and vector number exceptions. Hardware support is via the IPL0-2* with the AVEC* and IPEND* signals. The supervisor/user model is maintained to handle these and other exceptions, although the processor exception model is different.

Five stack frame formats are used for the different exceptions. The MC68040 uses an instruction restart exception model to handle bus transfer errors — sufficient information is stored in the 30 word stack frame to allow safe system continuation. This has several advantages, in that it reduces the exception stackframe size needed to store the internal contents and allows software and, therefore, the system designer, to decide how the system will handle such interruptions to normal processing. It is up to the exception handler to decide whether outstanding memory accesses are to be completed, for example.

Bus snooping

The bus snooping protocol used on the MC68040 is different from that of the MC88100 by virtue of how the stale data is updated. With the M88000, the attempted access is snooped, aborted, the main memory updated and the original access automatically restarted.

With the MC68040, the access is snooped but the processor decides whether to act as the memory block or allow the memory to be enabled. There are two snoop control pins, SC0* and SC1*, which control exactly how the processor responds.

Typically, these are connected to the user attribute pins which allows software to control which mechanism to use via the page descriptors. The MI* output allows the processor to inhibit the external memory so it can act as memory and supply or accept the correct data as determined by the

SC*0–1 coding. Snooping can be used with copyback or writethrough modes, depending on the number of external masters and their caching capabilities.

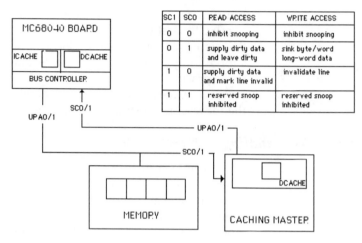

SC1	SC0	READ ACCESS	WRITE ACCESS
0	0	inhibit snooping	inhibit snooping
0	1	supply dirty data and leave dirty	sink byte/word long-word data
1	0	supply dirty data and mark line invalid	invalidate line
1	1	reserved snoop inhibited	reserved snoop inhibited

The MC68040 bus snooping

Meanwhile, RISC designs were also not holding back and were exploring techniques to execute multiple instructions per clock.

Superscalar alternatives

The first commercial RISC microprocessor to support some level of superscalar operation appeared from Intel. Its i860 processor allowed under some circumstances, the simultaneous execution of both an integer and a floating point instruction. While this facility did indeed offer multiple instructions per clock and high MIPS figures (100 MIPS from a 50 MHz processor), its implementation limited the amount of use that could be made of it.

The dual execution was controlled not by the hardware but by the compiler and software technology. To use this mode, an integer and floating point instruction would be linked to create an instruction pair. This was done by setting a bit within the instruction fields. The double instruction would then be executed simultaneously with each half being dispatched to the appropriate execution unit.

The disadvantages of this method are concerned with the compiler's ability to predict when it would be safe to execute such instruction pairs and in re-ordering the instruction flows to create them. The processor relies on the compiler to resolve conflicts by separating the conflicting instructions with others either through re-ordering or in the last resort, by inserting no-ops or preventing dual execution. These techniques proved extremely difficult for compiler technology to master, especially as the compiler had to err on the conservative side to preserve the processor's integrity. As a result, the opportunities offered within a program to exploit this facility were frequently small and thus there was often a large difference between the peak and the sustained performances.

Superpipelining

The MIPS R4000 processor also offers the ability to execute multiple instructions per clock but achieves this performance by using the superpipelining technique. With superpipelining, the instruction pipeline is double pumped to execute twice as many instructions and in this way can achieve a performance level that is equivalent to a two-instruction superscalar design.

Its advantages when compared to similar performance superscalar designs are due to simpler design and fabrication. A superscalar design will consume more transistor by replicating the execution units. A superpipeline design uses less transistors but relies on being able to clock the pipeline at double the clock speed. If silicon real estate is at a premium and the design can be double clocked, then superpipelining has a lot going for it.

The main problem faced with this type of design is in developing superpipelined processors that execute three or four instructions per clock. To do this requires the pipeline to be triple or quadruple pumped and this is often beyond the silicon fabrication technology. It is not clear how much the technology can be superpipelined. The history of CMOS technology has shown that transistor densities have increased far quickly than processor speeds. For example, the number of transistors in the MC68020 compared to the MC68060 give a density improvement ratio of about 20 times. The similar ratio for clock speeds is only about 5 times.

Overdrive chips

Superpipelining should not be confused with the overdrive technology that is so popular in today's PC processors. The overdrive technology as used with the Intel DX2 and DX4 80486 processor variants does involve running the processor and its pipelines at a far higher clock speed and if compared to the outside bus speed, the processor could be interpreted as executing multiple instructions per clock. In reality, the processors are running with a slow external bus as opposed to a superpipelined design. With superpipelining, multiple instructions are dispatched per clock and they are processed by double pumping the pipelines. The rest of the processor does not run at the pipelined clock rate. With an overdrive design, single instructions are dispatched every clock. When an external memory access occurs, the processor then slows down the external bus clock to at least half the internal speed.

Overdrive chips do offer speed improvements without having to change external bus circuitry but the improvements are only obtained when the instruction and data flow does not need to use the external bus. As soon as this happens, the faster processing is halted and performance is lost. Typically, an improvement of 1.5 to 1.6 is obtained by fitting a DX2 processor to a DX board.

Very long instruction word

The very long instruction word processor (VLIW) uses compiler technology to create an instruction word that comprises of multiple instructions. This very long word is then split into the components which are then issued in parallel to multiple execution units. This may seem to be very similar to a superscalar unit and this is a fair comment.

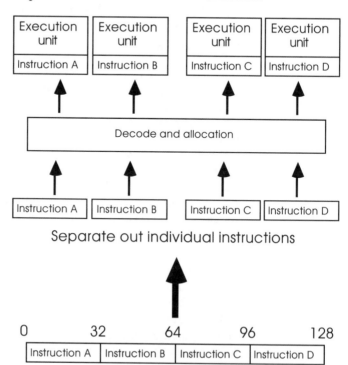

A VLIW design processor

The difference is where the responsibility lies for creating the very long word and thus controlling the multiple issue. With the VLIW approach, each field is effectively reserved for a specific execution unit and the compiler is responsible for filling the fields.

To do this, it will often unroll loops to create longer instruction sequences that it can use to fill the VLIW fields. As a result, there is trmendous code expansion. In addition, if the software does not exhibit much parallelism, it can be very difficult for the compiler to maintain and continually exploit the VLIW performance simply because the number of instructions it can use are limited either by the amount of parallelism or deficiencies in its scheduling algorithms.

With superscalar techniques, the internal hardware design is responsible, finally, in determining how much superscalar functioning is possible and does not need to rely on the compiler so much. However, this does require a more sophisticated decode logic compared to the VLIW approach.

VLIW had a brief moment of fame in 1990 when Philips presented a VLIW design at the International Solid State Circuits conference. The 200 bit design had five execution units and thus allowed the execution of a branch instruction, a register file access, two arithmetic operations and a memory access per clock. It was claimed that novel scheduling techniques were used in the compiler to schedule the instructions.

Interest in this approach has been rekindled by statements that Intel, Hewlett-Packard, IBM and Motorola are looking at this type of architecture. Whether this is the start of the replacement to RISC will remain to be seen.

Superscalar principles

Superscalar processors should not be confused with multiprocessor or parallel processing designs, although they are very similar in that they have the capability of executing multiple instruction every clock. The main difference is the level at which the allocation of resources is performed. With most parallel or multiple processor computers, the execution units virtually act independently and can be likened to several computer systems that simply reside in a single chassis with a single controlling program. The system software controls resource allocation, which may or may not be transparent to the user — applications may have to be adapted accordingly. The superscalar design performs this role at the VLSI hardware level so that the resource allocation is transparent to the user, thus enabling binary compatibility across a range of machines, irrespective of the number of instructions that can be executed per clock.

The general principles behind such a processor are relatively simple. The execution unit is replicated, so there are multiple units, each of which is capable of executing an instruction per clock cycle. These units are fed via an extremely wide bus (an eight unit design with a 32 bit op code needs a 256 bit bus), so that a very large word, a multiple instruction, can be presented.

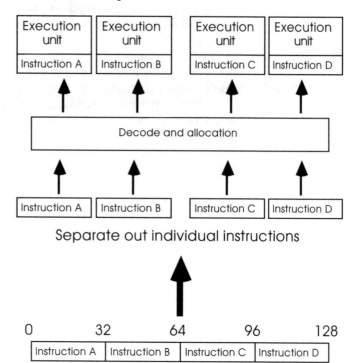

A 4 execution unit design

Each individual op code within it is allocated to a particular execution unit on every clock cycle. For efficiency, this requires a fixed op code length with no additional operands so single cycle memory accesses can be maintained and hardware needed to detect an op code overspill into the next allocated slot is removed. Fortunately, most RISC designs, including both the M88000 and PowerPC architectures, use a fixed size op code.

The next problem to overcome concerns the validity of the results. Although eight instructions may have been executed, not all the results are necessarily valid. If a branch instruction is encountered and taken, the instructions executed with it but located after it would not normally have been executed and their results should be discarded. Similarly, instructions which use results from previous instructions may not be valid because the data was not ready when they were executed. It is this type of problem that must be overcome before multiple execution machines can realise their potential performance.

Data dependency refers not only to data or registers encoded within the op code but also to any unique resource within the programming model, such as condition codes or status bits. If a programming model has a single condition code register, this immediately places restrictions on which instructions can be executed in parallel. Only one such instruction can be executed per clock, even though there may be tens of execution units available. In addition, there can be no automatic updating of the condition code bits as in the M68000 family — which instruction should be used for the updating? Architectures with dedicated condition code registers are not ideal for multiple execution units.

Conventional processors have a condition code register which is usually updated at the end of every instruction and allows easy condition testing. Various combinations of bits within the register indicate if certain conditions exist. Maintaining this register imposes a large overhead on the processor and it is no surprise that neither the M88000 family or the PowerPC have a specific condition code register. Instead, several registers are available to ensure that an alternative is available. With the M88000 family, any general purpose register can be temporarily used as a condition code register. With the PowerPC architecture, there are eight available.

Controlling multiple instructions per clock

The real difficulty in controlling a multiple instruction execution machine is removing or inhibiting the invalid data produced at the end of every clock cycle. There are two basic approaches — prevent the execution of the offending instruction so the erroneous data is not produced or remove it after execution. To do so after the event is extremely complicated and would require many buffers to act as temporary stores while the execution units progress the instructions. Each unit would need buffers to hold registers, a copy of the instruction, and temporary variables. Their contents would then be inspected, decoded and the register files etc. updated.

This job is even more complicated if the pipelines are then considered and are of differing lengths, as may be the case with a design with floating point units, integer units etc. With pipelines of equal length, the execution units can present their buffered data at the same time and thus provide an accurate snapshot of the processor on every clock cycle. The processing is simply an extra common stage within the pipeline.

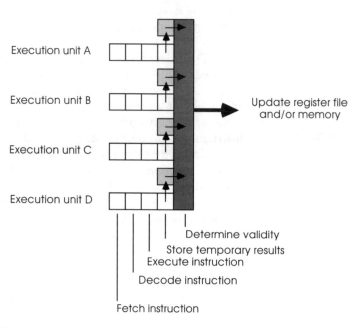

A post-event hard-ware mechanism for equal length pipelines

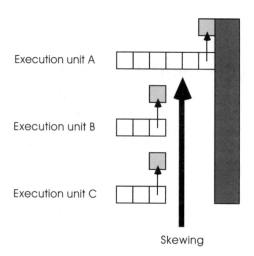

A post event hard-ware mechanism for different length pipelines

For unequal length pipelines, the temporary data cannot be made available within the same clock cycle and is skewed. The snapshot has to be delayed until unit A has stored its data. To preserve the temporary data from units B and C, their pipelines must either be stalled or the temporary data stored in a three stage FIFO buffer, allowing execution to continue. In this case, the pipelines have effectively been lengthened to that of unit A, which can cause further inefficiencies with branching, etc.

While such schemes can be implemented, the hardware is complicated and a better alternative is to prevent the production of invalid data in the first place.

Software control

The software techniques used to synchronise pipelines can be adapted to cope with multiple execution machines, as depicted. The machine shown has four execution units which each receive one instruction per clock. The first and second multiple instructions each contain a pair of data-dependent instructions. This dependency can be removed by reorganising, as previously described. The last instruction of the first pair and the first instruction of the second pair are interchanged, effectively delaying the former and advancing the latter. However, this does not take into account the potential problem concerned with the execution unit pipelines. Assuming that they are three stages deep, the pairs of dependant instructions need to be separated by at least another instruction and as a result, NOPs or other instructions are inserted accordingly.

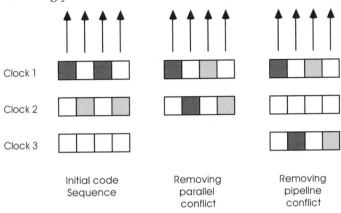

| | Initial code Sequence | Removing parallel conflict | Removing pipeline conflict |

n.b. shaded boxes indicate data dependant instructions

Software reorganisation to remove data dependency

The MC88110

The MC88110 was Motorola's second generation RISC processor and the first to use superscalar and out-of-order execution techniques to run multiple instructions per clock. The processor, fabricated in 0.8 μm CMOS using 1.3 million transistors with triple layer metal and double layer polysilicon, comprises of ten fully interlocked execution units — two integer, one bit field, two graphics and three for floating point addition, multiplication and division.

The instruction set is a superset of its predecessor, the MC88100, and includes pixel manipulation instructions — addition, subtraction, compare, packing, and rotation — to support the new graphics execution units. The floating point units can now access two register files — the standard MC88100 general register file and a second extended register file comprising of thirty two 80 bit registers for additional working space. They implement IEEE-754 compliant floating point but can also work in a time critical mode where a sensible result is delivered rather than trapping in the event of over and under flows and so on. This speeds up calculations and is beneficial in DSP and graphics applications.

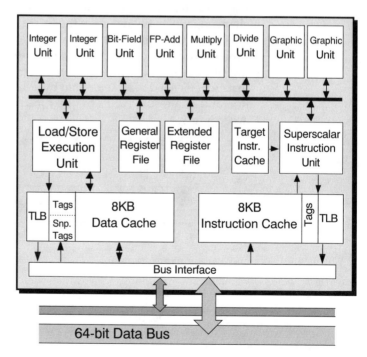

The MC88110 internal architecture

Symmetric superscalar

By virtue of its many execution units, the superscalar instruction issue is not restricted to a single floating point and integer instruction but can process about 89% of all instruction pairs — two integer, two floating point, graphics and integer and so on. There is no artificial restriction on how the pairs are organised. On each clock edge, the sequencer fetches two instructions from the instruction or branch target cache, decodes the instructions and fetches the necessary operands.

If the execution units are free and there is no data dependency, both instructions are issued. The sequencer always dispatches code in strict program order and if the first instruction of the pair cannot be issued, the second instruction is not issued either. The sequencer will wait until the dependency has been cleared before continuing. If the first instruction can be issued but the second cannot, the first instruction is issued, the second moved into the first instruction position and a new instruction fetched. In this way, the sequencer will issue as many instructions as possible while maintaining program order.

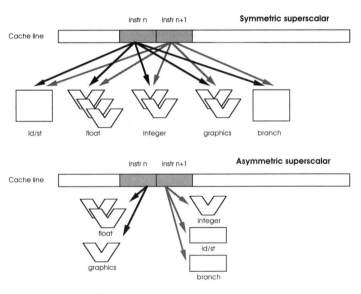

Symmetric and asymmetric superscalar operation

Once issued, the instructions are processed but their completion, i.e the writeback phase where registers are updated can complete out of order. This phase is again controlled by the sequencer which ensures that out-of-order completion does not affect data integrity. As with the MC88100, there is a feed forwarding mechanism where data can be fed to another pipeline while the writeback is performed, thus saving a clock cycle.

While compiler technology plays an important part in minimising data dependency and the occasions when dual issue is not possible, the processor can dynamically re-order the code internally to improve performance. If a store instruction cannot complete because of slow memory or other delay, it can be removed from the instruction pipeline to allow processing to continue and the store is completed at a later date in parallel with normal execution. When a branch is encountered, the processor can speculatively execute down one path of the branch, instead of waiting while the branch condition is evaluated. If the processor guesses right, execution simply proceeds as if nothing had happened. If it guesses wrong, the processor simply goes back to the branch and proceeds along the correct path. The direction the processor takes is determined by the type of branch instruction. In both these cases, the out-of-order execution is automatically halted if there is any data dependency or accesses to external memory which would destroy the system integrity.

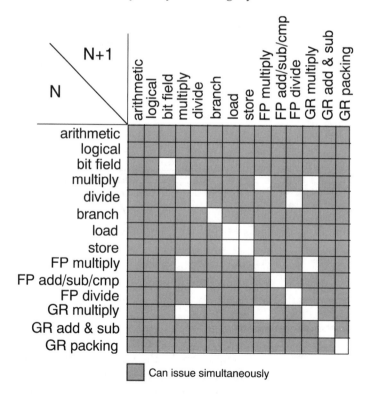

Dual execution modes

Although the device has a design frequency of 50 MHz, it has been designed to achieve high performance with a variety of memory configurations and the memory interface is very intelligent. The execution units are fed from two internal 8 Kbyte physically addressed caches, arranged in a Harvard architecture, that use the external bus to communicate to either DRAM or FSRAM or a secondary cache. The interface uses independent 64 bit data and 32 bit address buses and a burst mode to achieve a two-clock initial access followed by three single cycle accesses to fill a cache line. Several techniques are used to improve the bus efficiency: the buses can be used to support split transactions where the processor and other bus masters can utilise the data and address buses independently. While the data bus is being used by master A, a second bus master can use the address bus without having to wait for the master A to complete its data transaction. The interface uses wrap around addressing where the first address is that of the required data and the address wraps around to fill the remainder of the cache line. The data is fed directly from the bus pins into the waiting execution units to further reduce delay, and data from the rest of the burst can also be streamed directly to the execution units if needed. These techniques dramatically reduce the performance penalty of a cache miss. Further enhancements to the cache include the ability under software control to flush or pre-load them in parallel with normal processing. This allows data to be fetched ahead of its processing while the processor is not using the external bus and is busy elsewhere.

Full support for multi-processing environments is provided: a four-state MESI cache coherency protocol is implemented using bus snooping on the external bus and special snooping signals to ensure data cache coherency, and fast cache flush and invalidate operations are available for operating system software support. Dual cache tags are used to remove bus snooping penalty. In addition, the designer has the choice of either using hardware table walking to replace memory management descriptors or using his own software to fetch the descriptors.

The device is packaged in a 299 pin cavity down pin grid array. Simulated results indicate a Drhystone 2.1 performance of over 100 VAX MIPs and a floating point SPEC performance of over 60 with DRAM and 75 with a secondary cache.

Although many recognised the M88000 architecture as being the best RISC architecture from a technology standpoint, it was not a great success in terms of its adoption within the computer industry. Its hopes were pinned on the processor's adoption by Apple as the processor behind the next generation Macintosh personal computers. These hopes for the M88000 proponents were dashed in 1991 by the announcement that Apple, IBM and Motorola had formed a partnership to develop and use processors based on a new processor architecture called PowerPC which was based on the IBM POWER architecture which was used in IBM's RS/6000 workstations.

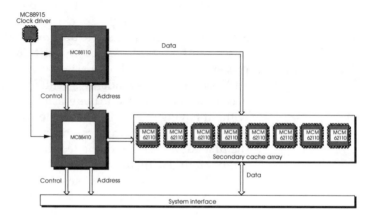

An example external cache configuration

For Motorola, it offered a second chance to dominate the RISC arena with the guarantee of the two dominant PC architectures adopting it. As could be expected, all further work on the M88000 family was stopped shortly after the announcement and the MC88110 was the last product to appear in the family.

Enter the PowerPC

The PowerPC architecture is the culmination of several previous IBM processor designs, starting with the IBM 801, which was the first IBM RISC implementation. The development of the IBM 801 led to the appearance of two further architectures: the short-lived RT-based PC and the more successful RS/6000 platform, which is also known as the POWER architecture. As shown in the diagram overleaf, the POWER architecture was the basis for the PowerPC architecture.

Although many of the POWER architecture instructions will execute on the PowerPC without modification, it is a mistake to assume that they are compatible. To allow RS/6000 software to run on a PowerPC processor, the unsupported instructions must be emulated in software. In this way, it is possible to have systems that are compatible with both the RS/6000 and PowerPC, without having complete binary compatibility at the processor level.

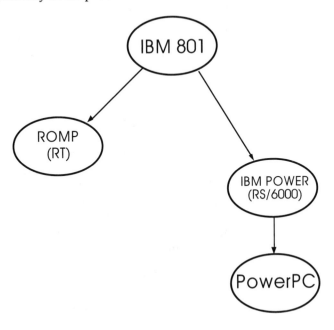

The PowerPC ancestry

The architecture, was not developed by IBM alone. It is the responsibility of the IBM–Motorola–Apple partnership which took the IBM POWER architecture as a foundation and derived the PowerPC architecture from it. Indeed, some of the architecture and particularly the hardware implementations, have used many of the ideas that appeared in Motorola's M88000 RISC architecture. For example, the bus interfaces of the MPC601 and MPC603 are similar to that found on the MC88110.

The PowerPC architectural model

The PowerPC model is a little different from the models put forward with previous processor families in that an architectural framework has been developed to encompass all family members from the beginning. When the Motorola MC68000 first appeared, it gave indications of how software should be implemented to allow compatibility with later generations, but the full details or techniques did not really appear until the new processors were released. In this way, the overall architectural model appeared to have evolved rather than be structured.

The PowerPC architecture is split into two sections: the architecture itself, which is common to PowerPC processors and provides a consistent programming interface and environment (including definitions of the 64 bit versions) and an implementation section, which is specific to each individual processor. Each section is itself is split into two parts.

This split roughly corresponds with the user and supervisor modes that PowerPC processors use to differentiate between the common and specific aspects of each processor's operation.

Each part is defined in an architectural book which describes the basic definitions. Book one covers the user instruction set, registers and logical address space. The second book describes the virtual storage model and includes the definitions on cache control. Together, they provide the basic architectural definition. The remaining two books are implementation specific. The third book covers the supervisor model and includes the basic exception handling and associated registers, while the fourth book defines the processor implementation details.

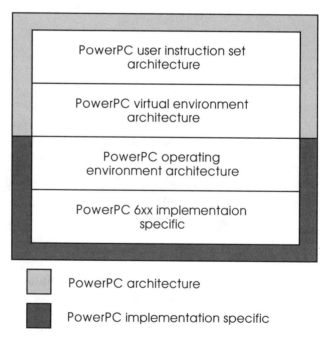

The PowerPC architectural model

The basic processor architecture, as defined in this documentation, is shown below. It consists of a branch unit, a fixed point or integer processing unit (which also contains the general purpose register file) a floating point unit and its own register file, and a data and instruction cache which communicates with memory.

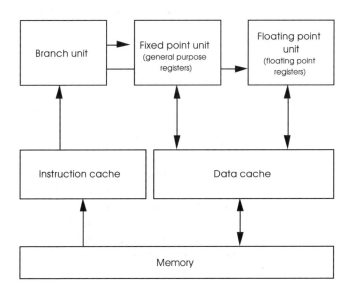

PowerPC architectural model

This architecture is capable of supporting superscalar operation, where multiple instructions can be executed per clock cycle. The PowerPC is almost unique in that all the processors currently available (i.e. the MPC601 and MPC603 and the previewed MPC604 and MPC620) are all superscalar devices.

Although the basic architecture does not appear to define it as so, and indicates the serial processing of instructions, the actual implementations process instructions internally in parallel and make use of out-of-order execution and storage. This appears to be a slight contradiction, however, the explanation is simple. The architectural documents define the programming model and environment that the software programmer will use to develop and write software. The actual implementation of the model is abstracted from this definition. Software does not need to know anything about the processor implementation, such as whether it is superscalar or capable of power down modes, and so on. This information is handled either by making it transparent, as in the case with superscalar operation, or passing it to the supervisor model, where all implementation-specific functions are handled. Powerdown management is handled in this way on the MPC603 by a special supervisor register.

Execution pipelines

The key to understanding the PowerPC processors is understanding how an instruction is processed using pipelines and knowing where the different responsibilities lie.

Basic theory

Pipelines have been used in both RISC and CISC processors for a long time to enable the complex task of fetching, decoding and executing an instruction to be split into several smaller sequential tasks. The instruction proceeds on every clock edge through the pipeline with a new instruction following behind it. This gives the effect of single cycle execution, even though the instructions take longer than a single clock to go through the pipeline.

In the example shown, the pipeline consists of four states: fetch, decode, execution and completion. An instruction enters and leaves the pipeline on every clock. On entering the pipeline, the instruction takes four clocks to complete, so the add instruction completes when the mul instruction enters the pipeline.

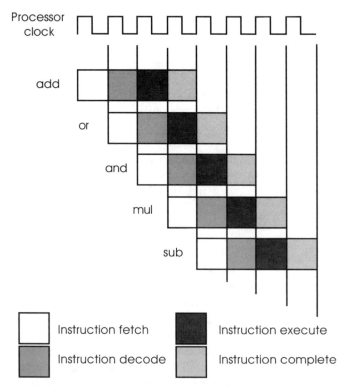

Example processor pipeline

PowerPC implementations

The working of the MPC601 and MPC603 execution pipelines is shown in the next diagram. This depicts a generic execution pipeline which could, for example, be an integer or floating point unit. It has four stages although, in many respects, the last stage is not really part of the pipeline. The reasoning behind this comment will become self-evident.

The first stage is to fetch the instruction. The instruction is received and the address of the next instruction calculated. The dispatch stage is responsible for decoding the instruction, getting the appropriate and passing the processed instruction to the execute stage. Here, the instruction is processed to provide the end result. This could take more than one cycle, in which case the data would loop around in this stage until the processing was complete. This would naturally stall the pipeline and prevent any other instructions from proceeding. This is the case with floating point divide instructions, for example, which take several clocks to process.

After execution, the result is passed to the final part — the complete/writeback stage. Once an instruction has been passed to this point, it has virtually completed. The previous three stages carry on processing and therefore, as far as they are concerned, the instruction has completed. However, all that has happened is that the results have been computed and have been passed on for storage.

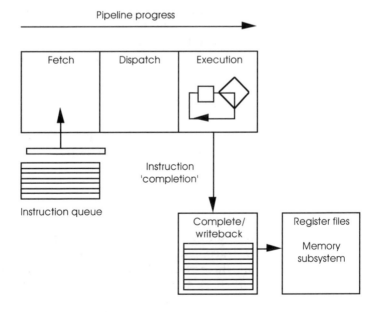

Execution unit pipeline

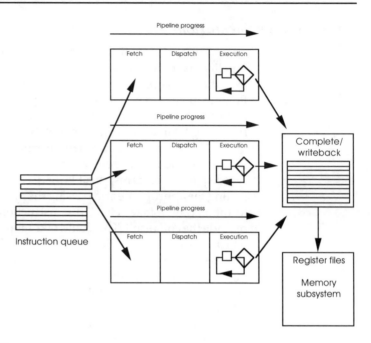

Going superscalar

This is analogous to an office worker giving a file to a clerk for filing: the office worker treats that as the completion but, in reality, the job is not finished until the file is safely put away. The complete/writeback stage performs the updating of registers and memory, as needed by the completing instructions when they come from the execution unit. This stage has some local storage to prevent the loss of data and stalling the pipeline if the information cannot be stored immediately. It has the added responsibility of ensuring that the instructions complete in the correct order. With a single pipeline, this is not a problem as the instruction flow received by the writeback stage will be the same as the program sequence because there is only one instruction stream. However, the PowerPC architecture is defined as superscalar and also supports out-of-order instruction execution; in these cases, this fourth stage has an important role to play.

Superscalar pipelines

The diagram shows a superscalar processor with three pipelines representing three independent execution units. Instructions are fed into each of these pipelines on every clock and three executed instructions are supplied to the complete/writeback unit.

The writeback unit now has an additional function to perform, i.e. making sure that the results are stored in the right order, as dictated by the program flow. The program flow may not be the same as the instruction flow to the three pipelines. To enhance performance, the pipelines may execute instructions speculatively by guessing the route a branch will go and executing down that line while the branch is being evaluated. Due to different processing times, an integer instruction that is after a floating point instruction may execute and finish before the floating point instruction finishes, and so on. (Both the MPC601 and the MPC603 use these techniques to improve performance.) As a result, the writeback unit has the responsibility for ensuring the correct machine context and for making sure that all data is modified and accessed in the correct program order, even though the instruction sequence within the processor was different — the appropriately named 'out-of-order' execution.

Data dependencies

The previous example assumed that all the instructions executed had no data dependencies, i.e. did not need the results of an instruction or a memory access before it was available. This is not always the case in real life and there must be a mechanism for synchronizing the processor if such dependencies occur. The next diagram shows an example of a synchronization problem.

The first instruction performs a floating point addition, where the contents of registers 2 and 3 are multiplied together and the result stored in register 1. The next instruction takes register 1, adds it to register 5 and stores the result in register 4. The problem occurs when the second instruction starts down the execution pipeline. The value of register 1 that it uses will not be the new value and the resulting calculation will be wrong. This would corrupt the system.

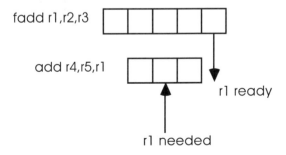

Pipeline synchronization problem

There are several solutions to this. The first relies on software to recognise all such situations and insert either instructions or no-ops to delay the second instruction.

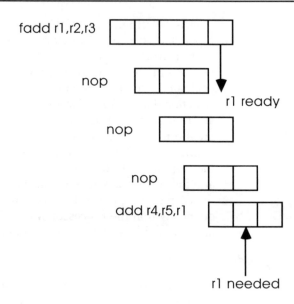

*A software solution
using NOPs*

This often wastes opportunities where instructions could have been executed and makes the entire hardware integrity totally dependent on software. If, for instance, the hardware changes with modifications to pipeline lengths, the software solution may no longer be valid and binary code compatibility is lost.

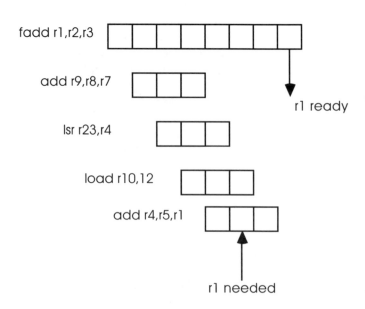

Processor corruption

The PowerPC architecture uses a hardware interlock to solve this type of problem. If a dependency is detected, the pipelines are stalled until it is resolved. This causes processing delays, which can be used by re-ordering the instruction sequence to use the otherwise lost processing time, but it does guarantee that the software and any instruction sequence will execute correctly on any PowerPC processor.

Handling exceptions

Up to now, the pipeline descriptions have assumed that the instruction execution does not get interrupted or suffer from any errors which would cause an exception condition for the processor to handle.

The PowerPC does not immediately take an exception when an instruction causes one. There are several reasons for this: first, the instruction may be executing out-of-order and thus may not actually need to be executed. In this case, the exception can be ignored. The second problem is again concerned with out of order execution: in this scenario, the exception cannot be ignored but the excepting instruction has executed ahead of other instructions. In this case, the processor must wait until these instructions have completed to ensure that the correct processor status is available to help resolve the exception.

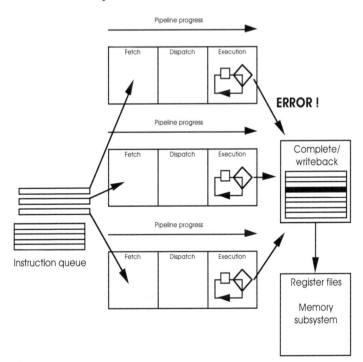

Handling errors

This is implemented by removing the offending instruction from the pipeline as soon as the exception is recognized and placing it into the writeback stage, marked as an exception causing instruction (marked black in the diagram). The writeback stage can then abort all instructions that follow it in the program flow and wait for all instructions that precede it to complete. The exception is then signalled as normal.

The advantage of this method is that the processor status is presented to the exception handler software in exactly the state as the processor would be in if no out-of-order or superscalar execution had occurred.

Branch delays

One major problem facing any processor architecture that executes instructions in a single cycle is how to handle changes in program flow caused by branch instructions without, for example, temporarily halting the program flow.

This difficulty is caused by the branch instruction not completing its operation before the next instruction is brought into the pipeline. If the branch instruction requires the program to move to a different address, this next instruction is not needed and must be removed from the pipeline. Further problems can be caused if the target instruction is not immediately available and has to be fetched from external memory. In short, every instruction that can change the program flow has the potential to significantly delay the instruction flow and execution.

To understand the problem, consider a processor that uses a three-stage pipeline as shown in the next two diagrams. The first diagram shows what happens when a branch instruction is introduced that does not change the program flow. If the branch is not taken, then the flow proceeds linearly and there is no execution delay. The resulting instruction flow is B – C – BRA – D.

In the second scenario, the branch is taken but the decision to go to the target address cannot be confirmed to cycle 3, by which time the next instruction D has entered the pipeline. This instruction should not be executed and must be removed. This occurs in cycle 4 and creates a hole in the pipeline which is often called a delay slot or bubble. This means that during the next cycle, no instruction is completed, and the branch instruction has become a two clock instruction. The resulting instruction flow is B – C – BRA – delay slot – K.

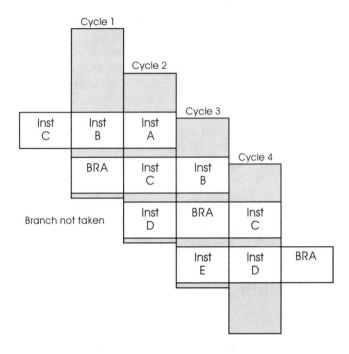

Branch not taken scenario

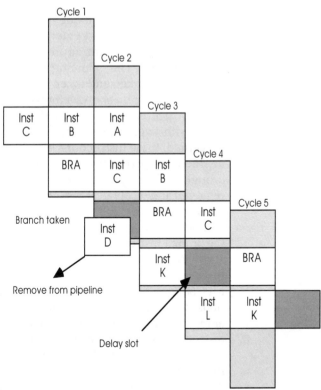

Branch taken scenario

This explanation assumes several characteristics about this type of pipeline: first, that the branch instruction is resolved when it has reached the second stage of the pipeline and secondly, that the target instructions are immediately available. In practice, these assumptions are not always valid. With a 'branch always' or jump type instruction, resolution is not dependent on values in other registers or the results of logical and arithmetic tests, and so can be resolved immediately. With conditional branches, the test results (or register contents that determine the test results) may not be ready and so the delay slot may occupy multiple cycles until such time as the situation is resolved.

Similarly, if the target instruction is not ready (i.e. if it has to be fetched from external memory or requires memory management intervention), this will also extend the delay slot time period.

Branch folding

Delay slots can be minimised by resolving branches as early as possible within the pipeline so that unwanted instructions are not introduced which require subsequent removal. The MPC601 and MPC603 processors both use this technique to improve branch performance.

The generic technique is simple: any branch instructions are removed as they enter the execution pipeline, immediately resolved and the appropriate instructions promoted into their place. The following two diagrams show how this works for not taken and taken branches.

The diagram shows a four-stage queue, where the top two instructions are dispatched to the processor execution units with every clock cycle and, as far as this queue is concerned, these instructions have completed. The next two instructions are then moved up and two new instructions are brought in. With the 'not taken' case, the branch instruction is removed and resolved at the same time as the top two instructions are completed. If the branch is not taken, the next three instructions are loaded and the execution flow continues as if the branch instruction was not in the code.

With the branch taken scenario, the three instructions that are brought in come from the target address. In both the MPC601 and MPC603 processors, there are sufficient instructions stored within the queue to promote forward and so the empty spaces are filled, irrespective of whether the branch instruction is first or second in the instruction pair.

This branch folding technique (so called because the flow changes are folded into the code so that it appears to be a single stream) only works if the branch can be immediately resolved. With conditional branches, this may not be the case and the processor's pipeline must be stalled until the condition can be resolved.

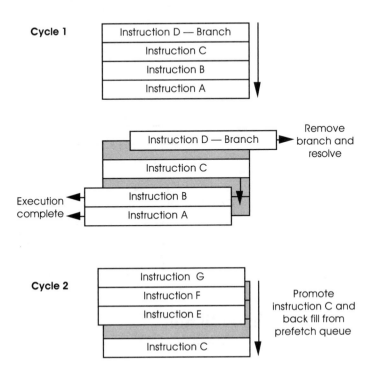

Branch folding —
branch not taken

If a condition code register is altered and this is not complete before the branch is encountered, the target address cannot be calculated and so the appropriate instructions cannot be inserted. If the link register is modified prior to its use by a branch instruction, the branch instruction must wait until the link register is updated. In such cases, the pipeline is again stalled.

With unconditional branches, there is less likelihood of a pipeline stall because the branch is always taken – but the execution pipeline can be stalled due to memory subsystem delays. If the instructions stored at the target address are not cached, and therefore require fetching from memory, there can be a considerable delay — and during this time the execution pipeline is stalled and cannot proceed. Strictly speaking, such delays are really classed as memory latency problems rather than branching difficulties, although the effect is the same: a stalled pipeline.

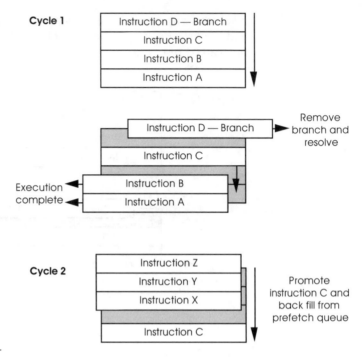

*Branch folding —
branch taken*

Static branch prediction

Both the MPC601 and MPC603 processors use an additional technique called 'static branch prediction' which helps to reduce the delays incurred when a branch instruction cannot be resolved. This method involves predicting which way the branch will go — whether it is taken or not — and processing that predicted instruction stream. When the branch is evaluated, the prediction can be verified and, if proved correct, processing just continues. If the prediction was wrong and this speculative execution is not part of the normal instruction flow, the processor backs ups and then continues down the correct instruction stream. The key is indicating to the branch execution unit which way it should go. The method used within the PowerPC family is to allow the software to indicate the prediction through the encoding within the branch instruction itself. The y bit is used to encode the prediction as follows:

- With y = 0 and the displacement is negative, i.e. the flow goes back, the prediction is that the branch is taken.
- With y = 0 and the branch uses either the count register or link register as the target destination, the prediction is that the branch is not taken.

- With y = 1 and the displacement is negative, i.e. the flow goes back, the prediction is that the branch is not taken.
- With y = 1 and the branch uses either the count register or link register as the target destination, the prediction is that the branch is taken.
- If no prediction can be made, y should be set to 0.

There is one important point to remember about these encodings: they are processor specific and may change with later versions of the PowerPC family, or even not be recognised or used at all!

The predicted instruction stream is executed as far as possible without modifying data either internally or externally. This means that the flow will stall if an instruction's processing reaches the writeback stage and needs to output to cache or external memory or an internal register. This typically limits the predicted instruction execution to only a few clocks but, in many cases, this provides sufficient time to resolve the branch condition.

Branch prediction cache

An alternative to static branch prediction is to cache the results of a branch so that the next time that it is taken, the result of the last time can be immediately used. This technique involves caching the instruction address and the target address. It gains performance by assuming that most branch instructions will repeat and thus the address taken on the first encounter will be the target address for the majority of times that the branch will execute. In this way, the penalty and performance delay suffered during the first execution are offset by gains for every other time that the branch repeats that particular flow. While the branch is used to continue the loop, this will be the case. When the loop is exited, the branch cache does not predict correctly and thus does not supply the target address. At this point, it no longer speeds up the process until the next time the branch instruction is executed, i.e. the loop is executed again. With the branch instrucion cached as a result of the previous loop execution, the new loop can take advantage of the cache entry and gain performance.

The problem with this technique is that of the cache hit ratio. The caches are typically small and, to reduce complexity and thus silicon, are frequently restricted to a small number of entries between 32 and 128 entries and use logical addresses. In tight loops, the cache can have a big advantage as this increases the hit ratio. In less constrained and tight code, the hit ratio is not so high and thus the performance decreases. With logically addressed branch caches, every context switch requires that the cache is flushed and this further reduces its efficiency.

This technique came to the fore with RISC devices when it appeared on the AMD29000 device. Its incorporation into other processors has really depended on the performance advantages versus the silicon area taken by the cache hardware. In many cases, the combination of static prediction and using the silicon area to increase the data and instruction caches or add execution units have provided a better performance gain and thus the decision has gone against the incorporation of branch prediction caches. However, it is starting to become a standard feature — it is included on the superscalar MC68060 — and is likely to be included on the next generation of PowerPC processors.

Register renaming

Register renaming is, as its name implies, a method of changing the contents of a register by changing its name. It allows a speculative instruction stream to continue processing, even if it modifies a register's contents.

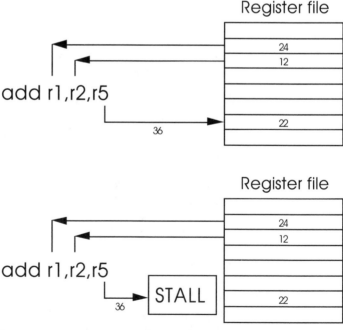

Speculative execution without register renaming

The normal problem with allowing a predicted instruction stream to proceed and modify a register or memory, for that matter, is concerned with the potential loss and corruption of data. If the register or memory is updated and the prediction is wrong, the original contents are lost, having been replaced with a now incorrect value.

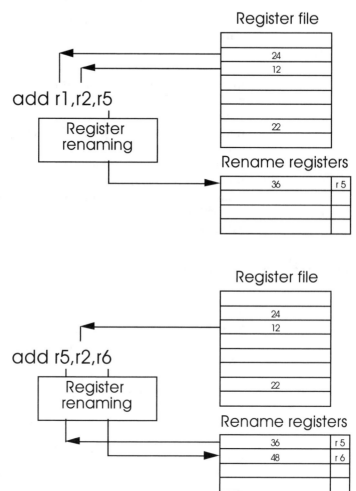

Speculative execution
with register renaming

With register renaming, the register modification is allowed to proceed, however, the value is not stored in the register but in an alternative register. In the example, the result of adding r1 to r2 (the value 36) is not placed in the register file, but is stored in the first rename register. The name of the destination register that it will replace is stored, together with the value. This preserves the original contents of the register. This alternative register then replaces the original register, as far as the predicted instruction stream is concerned, by re-naming it as the original register. The second instruction in the example needs to add the value or r5 to r2 and gets the value of r2 from the register file — but r5 comes from the renamed register that is replacing it. As before, the result must be stored in a rename register to prevent loss of the contents of r6, and so is stored in the next rename register.

If the speculation is incorrect, the original register is renamed and the original contents restored. Needless to say, all the modifications to the alternative registers are ignored.

By expanding the nomenclature, it is possible to use this technique to cope with multiple speculative paths which could occur if the instruction stream has several branch instructions in succession, or jumps to other branch instructions. In these cases, tagging the alternative registers with speculative path identities, the processor can execute down several pathways simultaneously and still be able to backtrack to the correct point, depending on the eventual branch evaluations. This advanced technique is not incorporated in the MPC601 or MPC603 processors but is include in the MPC604.

The MPC601 block diagram

The MPC601 was the first PowerPC processor available. It has three execution units: a branch unit to resolve branch instructions, an integer unit and a floating point unit.

The floating point unit supports IEEE format. The processor is superscalar. It can dispatch up to two instructions and process three every clock cycle. Running at 66 MHz, this gives a peak performance of 132 million instructions per second.

The branch unit supports both branch folding and speculative execution where the processor speculates which way the program flow will go when a branch instruction is encountered and start executing down that route while the branch instruction is resolved.

The general purpose register file consists of 32 separate registers, each 32 bits wide. The floating point register file also contains 32 registers, each 64 bits wide, to support double precision floating point. The external physical memory map is a 32 bit address linear organisation and is 4 Gbytes in size.

The MPC601's memory subsystem consists of a unified memory management unit and onchip cache which communicates to external memory via a 32 bit address bus and a 64 bit data bus. At its peak, this bus can fetch two instruction per clock or 64 bits of data. It also supports split transactions, where the address bus can be used independently and simultaneously with the data bus to improve its utilisation. Bus snooping is also provided to ensure cache coherency with external memory.

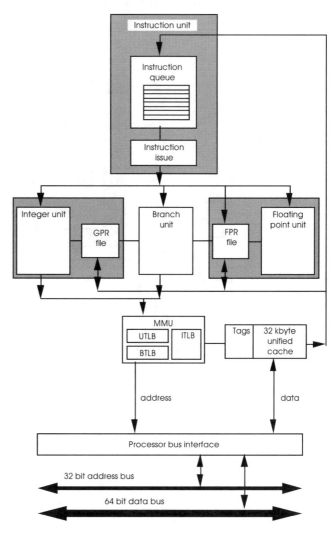

MPC601 internal block diagram

The cache is 32 kbytes and supports both data and instruction accesses. It is accessed in parallel with any memory management translation. To speed up the translation process, the memory management unit keeps translation information in one of three translation lookaside buffers.

The MPC603 block diagram

The MPC603 was the second PowerPC processor to appear. Like the MPC601, it has the three execution units: a branch unit to resolve branch instructions, an integer unit and a floating point unit.

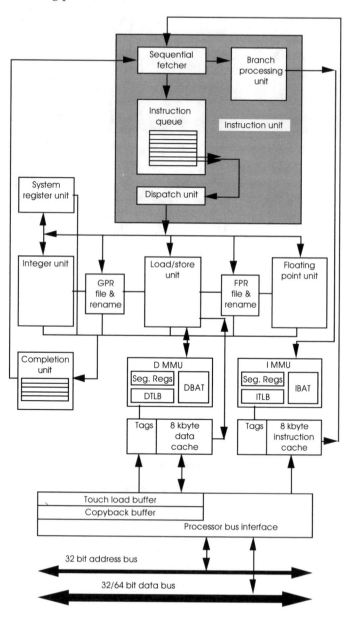

MPC603 internal block diagram

The floating point unit supports IEEE format. However, two additional execution units have been added to provide dedicated support for system registers and to move data between the register files and the two onchip caches. The processor is superscalar and can dispatch up to three instruction and process five every clock cycle.

The branch unit supports both branch folding and speculative execution. It augments this with register renaming, which allows speculative execution to continue further than allowed on the MPC601 and thus increase the processing advantages of the processor.

The general purpose register file consists of 32 separate registers, each 32 bits wide. The floating point register file contains 32 registers, each 64 bits wide to support double precision floating point. The external physical memory map is a 32 bit address linear organisation and is 4 Gbytes in size.

The MPC603's memory subsystem consists of a separate memory management unit and on chip cache for data and instructions which communicates to external memory via a 32 bit address bus and a 64 or 32 bit data bus. This bus can, at its peak, fetch two instruction per clock or 64 bits of data. Each cache is 8 Kbytes in size, giving a combined on chip cache size of 16 Kbytes. The bus also supports split transactions, where the address bus can be used independently and simultaneously with the data bus to improve its utilisation. Bus snooping is also provided to ensure cache coherency with external memory.

As with the MPC601, the MPC603 speeds up the address translation process, by keeping translation information in one of four translation lookaside buffers, each of which is divided into two pairs, one for data accesses and the other for instruction fetches. It is different from the MPC601 in that translation tablewalks are performed in software and not automatically by the processor.

The device also includes power management facilities and is eminently suitable for low power applications.

5　　Digital signal processors

When confronted with the term digital signal processing (DSP), an immediate response is an image of complex mathematical formula requiring banks of computers to perform even trivial tasks as filtering out high frequency signals from low frequency ones. This tends to beg the question, why bother?

The 'mathematical complexity' is attributable to years of experience and familiarity with analogue filters and their relatively simple equations. A simple filter is extremely simple compared to its digital equivalent. The analogue filter works by varying the gain of the operational amplifier which is determined by the relationship between r_i and f.

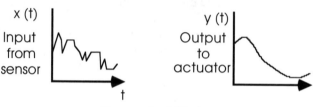

The required filtering

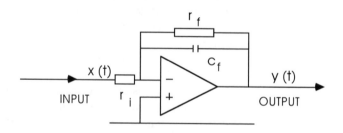

The analogue circuit

$$\frac{y(t)}{x(t)} = -\frac{r_f}{r_i}\left[\frac{1}{1 + j\omega\, r_f c_f}\right]$$

The mathmatical function

Analogue signal processing

In a system with no frequency component, the capacitor c_f plays no part as its impedance is far greater than that of r_f. As the frequency component increases, the capacitor impedance decreases until it is about equal with r_f where the effect will be to reduce the gain of the system. As a result, the amplifier acts as a low pass filter where high frequencies will be filtered out. The equation shows the relationship where $j\omega$ is the frequency component. These filters are easy to design and are cheap to build. By making the CR network more complex, different filters can be designed.

The digital equivalent is more complex requiring several electronic stages to convert the data, process it and reconstitute the data. The equation appears to be more involved, comprising of a summation of a range of calculations using sample data multiplied by a constant term. These constants take the place of the CR components in the analogue system and will define the filter´s transfer function. With digital designs, it is the tables of coefficients that are dynamically modified to create the different filter characteristics.

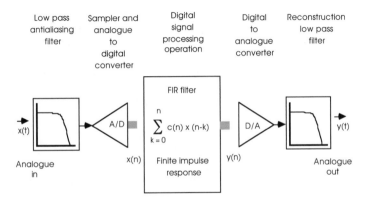

| Low pass antialiasing filter | Sampler and analogue to digital converter | Digital signal processing operation | Digital to analogue converter | Reconstruction low pass filter |

Digital signal processing

Given the complexity of digital processing, why then use it ? The advantages are many. Digital processing does not suffer from component ageing, drift or any adjustments which can plague an analogue design. They have high noise immunity and power supply rejection and due to the embedded processor can easily provide self-test features. The ability to dynamically modify the coefficients and therefore the filter characteristics allows complex filters and other functions to be easily implemented. However, the processing power needed to complete the 'multiply-accumulate' processing of the data does pose some interesting processing requirements. The diagram shows the problem. An analogue signal is sampled at a frequency f_S and is converted by the A/D converter. This frequency will be first determined by the speed of this conversion. Every period, t_S, there will be a new sample to process using N instructions. The table shows the relationship between sampling speed, the number of instructions and the instruction execution time. It shows that the faster the sampling frequency, the more processing power is needed. To achieve the 1 MHz frequency, a 10 MIPS processor is needed whose instruction set is powerful enough to complete the processing in under 10 instructions. This analysis does not take into account A/D conversion delays. For DSP algorithms, the sampling speed is usually twice the frequency of the highest frequency signal being processed: in this case the 1 MHz sample rate would be adequate for signals up to 500 kHz. Certainly for higher frequencies up to radio frequencies, analogue filters still have a lot to offer.

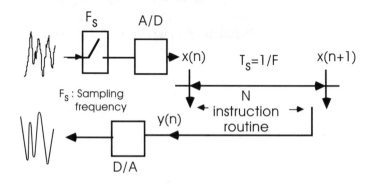

Instruction cycle	F_s	T_s	No. of instructions between two samples
1 µs	1khz	1ms	1k
	10khz	100µs	100
	100khz	10µs	10
	1Mhz	1µs	1
100ns	1khz	1ms	10k
	10khz	100µs	1k
	100khz	10µs	100
	1Mhz	1µs	10

DSP processing requirements

The analysis also highlights one major criterion that a DSP processor must fulfil. It shows that to maintain the periodic sampling, necessary to stop distortion and maintain the algorithm´s integrity, the processing must be completed within a finite number of instructions. This immediately raises the question of how to handle interrupts, program loops, etc. Most of the processor enhancements are centred around this issue.

One major difference between analogue and digital filters is the accuracy and resolution that they offer. Analogue signals may have definite limits in their range, but have infinite values between that range. Digital signal processors are forced to represent these infinite variations within a finite number of steps determined by the number of bits in the word. With an 8 bit word, the increases are in steps of 1/256 of the range. With a 16 bit word, such steps are in 1/65,536 and so on. Depicted graphically as shown, a 16 bit word would enable a low pass filter with a roll off of about 90 dB. A 24 bit word would allow about 120 dB roll off to be achieved.

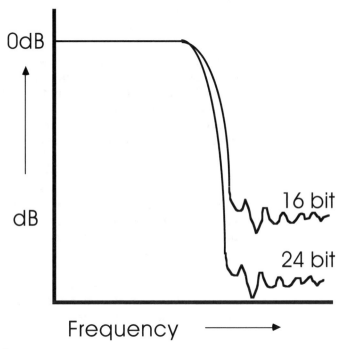

Word size and cutoff frequencies

DSP can be performed by ordinary microprocessors, although their more general purpose nature often limits performance and the frequency response. However, with responses of only a few hundred Hertz, even simple microcontrollers can perform such tasks. As silicon technology improved, special building blocks appeared allowing digital signal processors to be developed, but their implementation was often geared to a hardware approach rather than designing a specific processor architecture for the job.

Processor requirements

A good example of a DSP function is the finite impulse response (FIR) filter. This involves setting up two tables, one containing sampled data and the other filter coefficients that determine the filter response. The program then performs a series of repeated multiply and accumulates using values from the tables. As shown, the bandwidth of such filters depends on the speed of these simple operations. With a general purpose architecture like the M68000 family the code structure would involve setting up two tables in external memory, with an address register allocated to each one to act as a pointer. The beginning and the end of the code would consist of the loop initialisation and control, leaving the multiply-accumulate operations for the central part. The M68000 instruction set does offer some facilities for efficient code: the incremental addressing allows pointers to progress down the tables automatically, and the decrement and branch instruction provides a good way of implementing the loop structures. However, the disadvantages are many: the multiply takes >40 clocks, the single bus is used for all the instruction fetches, and table searches, thus consuming time and bandwidth. In addition the loop control timings vary depending on whether the branch is taken or not. This can make bandwidth predictions difficult to calculate. This results in very low bandwidths and therefore of limited use within digital signal processing. This does not mean that a MC68000 cannot perform such functions: it can, providing performance is not of an issue.

RISC architectures like the M88000 family can offer some immediate improvements. The MC88100´s capability to perform single cycle arithmetic is an obvious advantage. The Harvard architecture reduces the execution time further by allowing simultaneous data and instruction fetches. The MC88100 can, by virtue of its high performance, achieve performances suitable for many DSP applications. The system cost is high involving a multiple chip solution with very fast memory etc. In applications that need high speed general processing as well, it can also be a suitable solution.

Another approach is to build a dedicated processor to perform certain algorithms. By using discrete building blocks, such as hardware multipliers, counters, etc., a totally hardware solution can be designed to perform such functions. Modulo counters can be used to form the loop structures and so on. The disadvantages are of cost and a loss of flexibility. Such hardware solutions are difficult to alter or program. What is obviously required is a processor whose architecture is enhanced specifically for DSP applications.

The DSP56000 family

When Motorola announced its DSP56000 family of DSP processors, it was a great surprise not only to the industry but also to most of the people within Motorola! It was not the first DSP chip, both Texas Instruments and NEC had been supplying chips in volume, but it was the first generation to take a classical microprocessor architecture and enhance it for DSP applications. It is this combination that has led it to be used in many embedded controller applications which, while not performing traditional DSP functions, can use its fast arithmetic processing, onchip peripherals and memory to great effect.

The first two members of the family were the DSP56000 which has a mask programmed internal program ROM and the DSP56001 which replaces the ROM with RAM. For most applications, they are completely interchangeable. There is now a faster processor called the DSP56002 which uses the same generic architecture and a fourth version called the DSP56004 which is designed for digital audio applications.

Basic architecture

The processor is split into ten functional blocks. It is a 24 bit data word processor to give increased resolution[1]. The device has an enhanced Harvard architecture with three separate external buses: one for program and X and Y memories for data. The communication between these and the outside world is controlled by two external bus switches, one for data and the other for addresses. Internally, these two switches are functionally reproduced by the internal data bus switch and the address arithmetic unit (AAU). The AAU contains 24 address registers in three banks of 8. These are used to reference data so that it can be easily fetched to maintain the data flow into the data ALU. The program address generator, decode controller and interrupt controller organise the instruction flow through the processor. There are six 24 bit registers for controlling loop counts, operating mode, stack manipulation and condition codes. The program counter is 24 bit although the upper 8 bits are only used for sign extension.

The main workhorse is the data ALU, which contains two 56 bit accumulators A and B which each consist of three smaller registers A0, A1,A2, B0,B1 and B2. The 56 bit value is

[1] Yes, Motorola does define different word lengths for its different processors! The MC68000 word is 16 bits, the M88000 word is 32 bits and the DSP56000 word is 24 bits. To avoid confusion, I have referred to 16 bit words, 24 bit word and so on.

stored with the most significant 24 bit word in A1 or B1, the least significant 24 bit word in A0 or B0 and the 8 bit extension word is stored in A2 or B2. The processor uses a 24 bit word which can provide a dynamic range of some 140 dB, while intermediate 56 bit results can extend this to 330 dB. In practice, the extension byte is used for over and underflow. In addition there are four 24 bit registers X1, X0, Y1 and Y0. These can also be paired to form two 48 bit registers X and Y.

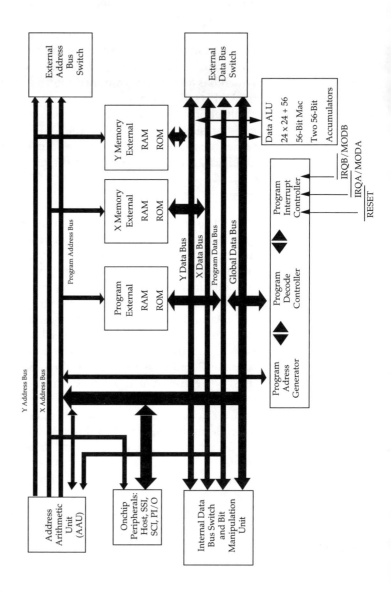

The DSP56000 block diagram

These registers can read or write data from their respective data buses and are the data sources for the multiply accumulate (MAC) operation. When the MAC instruction is executed, two 24 bit values from X0,X1,Y1 or Y0 are multiplied together, and then added or subtracted from either accumulator A or B. This takes place in a single machine cycle of 75 ns at 27 MHz. While this is executing, two parallel data moves can take place to update the X and Y registers with the next values. In reality, four separate operations are taking place concurrently.

The data ALU also contains two data shifters for bit manipulation and to provide dynamic scaling of fixed point data without modifying the original program code by simply programming the scaling mode bits. The limiters are used to reduce any arithmetic errors due to overflow for example. If overflow occurs, i.e. the resultant value requires more bits to describe it than are available, then it is more accurate to write the maximum valid number than the overflowed value. This maximum or limited value is substituted by the data limiter in such cases, and sets a flag in the condition code register to indicate what has happened.

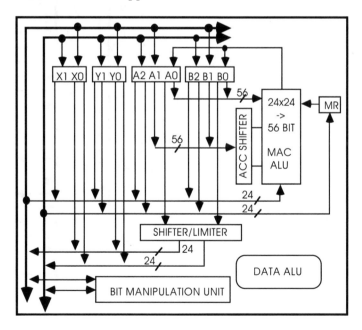

The DSP56000 data ALU

The external signals are split into various groups. There are three ports A, B and C and 7 special bus control signals, two interrupt pins, reset, power and ground and finally, clock signals. The device is very similar in design to an 8 bit microcontroller unit (MCU), and it can be set into several different memory configurations.

The three independent memory spaces, X data , Y data and program are configured by the MB,MA and DE bits in the operating mode register. The MB and MA bits are set according to the status of the MB and MA pins during the processor´s reset sequence. These pins are subsequently used for external interrupts. Within the program space, the MA and MB bits determine where the program memory is and where the reset starting address is located. The DE bit either effectively enables or disables internal data ROMs which contain a set of Mu and A Law expansion tables in the X data ROM and a four quadrant sine wave table in the Y data ROM. The onchip peripherals are mapped into the X data space between $FFC0 and $FFFF. Each of the three spaces is 64 Kbytes in size.

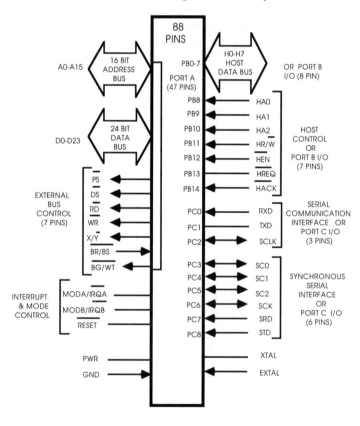

*The DSP56000/1
external pinout*

These memory spaces communicate to the outside world via a shared 16 bit address bus and a 24 bit data bus. Two additional signals, PS* and X/Y* identify which type of access is taking place. The DSP56000 can be programmed to insert a fixed number of wait states on external accesses for each memory space and I/O. Alternatively, a asynchronous handshake can be adopted by using the bus strobe and wait pins (BS* and WT*).

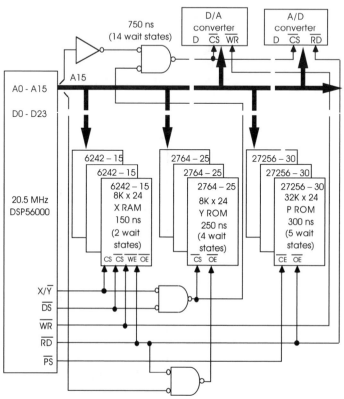

A mixed speed memory subsystem

There is a special bootstrap mode which allows the DSP56001 to copy its program from a single external EPROM into its internal RAM. Once loaded, the program is executed from internal memory without any speed penalty. This is a very common technique with MCUs, and again shows how the DSP56000 family has taken many MCU ideas and used them.

External communication can be performed via three ports A, B and C. Port A is used, together with the external bus control signals, to provide the expansion buses to communicate with external memory. Port B has 15 pins which can be configured either as general purpose I/O or as a byte wide full duplex parallel port, suitable for connection to the data bus of another processor. Port C has 9 pins which can also be configured as general purpose I/O or as two independent serial interfaces. The serial communication interface (SCI) is a three wire, full duplex serial communication port suitable for linking to other DSP processors, microprocessors, modems, etc. It can be used directly or with RS232C type transceivers. The three signals transmit data, receive data and generate a reference serial clock. It will support standard asynchronous bit rates or a 2.5 Mbit/s synchronous transfer. A programmable baud rate generator is used to generate the transmit and receive clocks or can be used as a timer if the SCI is not used. Multidrop mode is supported within the asynchronous protocols to allow master/slave operation.

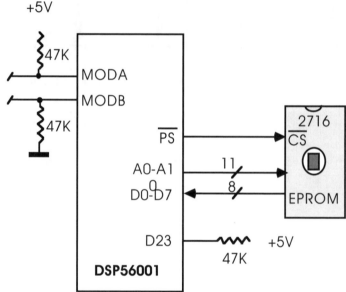

Bootstrapping from a single EPROM

A synchronous serial communication interface (SSI) is also provided to communicate to a variety of serial devices like codecs (a combined A/D and D/A used in digital telephone systems), other DSP processors, microcontrollers, etc., either on a periodic basis within a network or on demand. The number of bits per word, the protocol or mode, the clock and the transmit/receive synchronisation can all be defined. The 'normal' mode is typically used to interface on a regular or periodic basis between devices. The on-demand mode is data driven where the communication is irregular. In this mode the clock can be gated or continuous. In the network mode uses time slots to effectively allow multiple devices to communicate. From 2 to 32 devices can be supported. Each time slot is allocated to a device and, either through an interrupt or by polling status bits, each device has the option of transmitting data or not. The receiver always reads the data, thus allowing the processor the choice of using or discarding the data. The SSI is often used within arrays of DSP56000 processors to create a multiprocessing array to improve throughput or resolution.

The host port provided on port C can be used for general I/O but is usually configured to provide a byte wide parallel interface between the DSP56000 and other processors. The interface simply looks like a location within its own memory map which is read or written to by any intelligent device including DMA. Using its controls signals, it is easy to connect to an MCU or M68000 processor as shown.

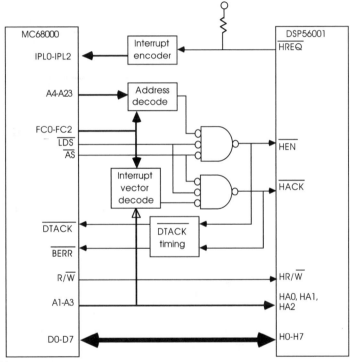

Using the HOST Interface

The host interface has a simple programming model. The three address bits HA0-HA2 allow access to the three transmit and receive byte registers, the command vector register and the interrupt control, vector and status registers. This can be programmed to allow either the DSP56000 or the other device to interrupt the other, thus allowing a fast, efficient data transfer. The HR/W* signal indicates the direction of data transfer and is used in conjunction with HEN* to enable transfers on the data bus.

The HREQ* signal is an open drain output signal used by the DSP56000 to request service from the other DMA device, processor, etc. It is normally used to generate an interrupt. The HACK* signal has two functions—it can either provide a host acknowledge handshake for DMA transfers or be used as an M68000 host interrupt acknowledge signal which places the contents of the interrupt vector register on the data bus to generate the interrupt vector.

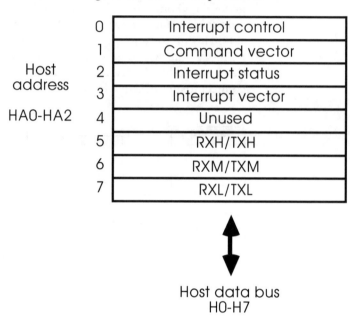

0	Interrupt control
1	Command vector
2	Interrupt status
3	Interrupt vector
4	Unused
5	RXH/TXH
6	RXM/TXM
7	RXL/TXL

Host address

HA0-HA2

Host data bus
H0-H7

The host interface registers

An extremely innovative feature of the interface is the host command facility. This allows the host processor to issue vectored exception requests direct to the DSP56000. The host can select one of 32 exception routines by writing into the command vector register to perform operations such as reading and writing to the DSP56000 registers, X and Y memory, force specific exception handlers for the onchip peripherals, perform control and debugging routines, etc. In some respects, this is similar to the MC68020 coprocessor interface which uses microcoded routines and the standard processor interface to allow the seamless joining of the main processor and the floating point unit. With the host command facility, the DSP56000 can be similarly coupled with another processor to combine both architectures into an extremely powerful tightly coupled multiprocessor system. This technique will be further explored in later chapters.

The programming model

The programming model is really a cross between the dedicated register model of the older 8 bit architectures and the general purpose, more orthogonal M68000 architecture. The register set can be divided into three main groups: the data ALU, the address generation unit and the program controller.

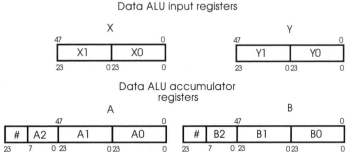

The DSP56000 programming model — data ALU unit

The data ALU register looks similar to those of an 8 bit architecture with its accumulators and X and Y registers, rather than that of a M68000. The dedicated registers come from the specialised hardware needed to perform the multiply-accumulate function. For DSP applications, the dual set is sufficient for most needs — and is still an improvement over a M6800 model.

The program generation set has two interesting components in addition to the normal program counter, status and operating mode registers. The loop address and loop counter registers allow hardware to perform do loops and instruction repetitions without any software overhead. Considering many routines are based around small instruction loops, this is extremely beneficial. The stack pointer is different from both the M68000 and M88000 families in that it is hardware based and capable of nesting up to 15 exception routines, before requiring software intervention to extend it into RAM. The 32 bit wide stack is only used to store the program counter and status register. Other registers are saved in either X or Y memory. In reality, DSP and embedded control software do not require to context switch so much and this is therefore not a disadvantage.

The address generation unit consists of three groups of eight registers, each of which provides an address pointer whose contents can be offset by the corresponding offset register and whose arithmetic function is defined by the related modifier register.

The pointer and offset registers behave in a similar way to the M68000 address registers, supporting the modes as shown. The different arithmetic address functions defined by the modifiers are for DSP support. The linear addressing mode is the standard twos complement 16 bit arithmetic function normally associated with microprocessors. The modulo addressing mode effectively causes the address contents to remain within a predefined address range with upper and lower boundaries. This is useful for repetitive table searching and for both general purpose processing and DSP applications.

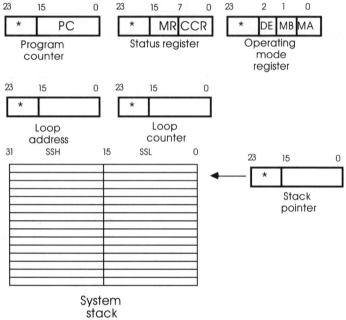

The DSP56000 programming model program controller

The reverse carry arithmetic is solely based on improving the execution of fast Fourier transforms (FFT). The FFT is a mathematical method of analysing a waveform into its component parts. Performing an FFT on an audio signal results in a frequency spectrum giving the intensities and frequency of the various components. The basic mathematical operation is a matrix multiplication on large arrays of data. By performing these calculations in a specific order, it is possible to reduce the number of calculations, thus increasing the speed or resolution of the analysis. The disadvantage is that the data order is scrambled and needs to be reordered, with the associated software and time overhead. The scrambling is not random and is equivalent to propagating the carry bit in the reverse direction. To save the software overhead, reverse carry arithmetic allows the data to be accessed in the correct order without reordering. In large FFTs, this is extremely advantageous.

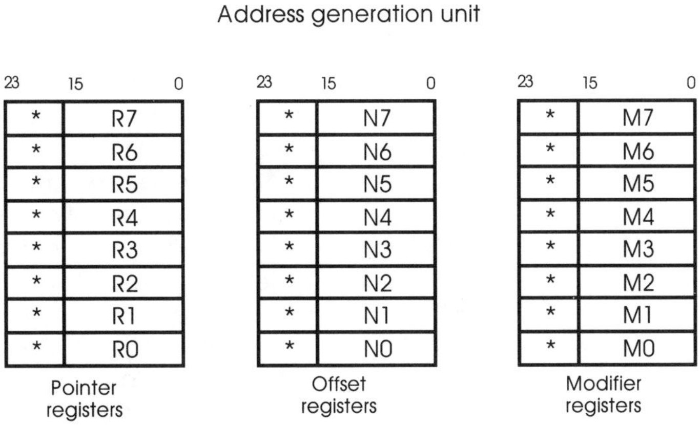

Address generation unit

23	15	0
*	R7	
*	R6	
*	R5	
*	R4	
*	R3	
*	R2	
*	R1	
*	R0	

Pointer registers

23	15	0
*	N7	
*	N6	
*	N5	
*	N4	
*	N3	
*	N2	
*	N1	
*	N0	

Offset registers

23	15	0
*	M7	
*	M6	
*	M5	
*	M4	
*	M3	
*	M2	
*	M1	
*	M0	

Modifier registers

The DSP56000 programmers model address generation

The instruction set

The instruction set looks like a hybrid between an M6800 8 bit processor and a M68000 16/32 bit chip. This follows from the programming model, which has dedicated M6800-like arithmetic accumulators with a set of M68000-like general purpose address registers. This hybridization permeates the instruction set where these types of op codes sit side by side. The set covers six main operation types:

- Arithmetic
- Logical
- Bit manipulation
- Loop
- Move
- Program control

Each instruction is coded within a 24 bit word.

Arithmetic and logical instructions

The arithmetic instructions perform all the arithmetic operations within the data ALU. They are register based and do not specify addresses on the X, Y or global data buses.

This reduces the number of bits needed to encode the operation, which then allows the spare bits to be used to specify parallel data moves across the X and Y space. Therefore, within most of the basic commands, the programmer can specify two additional transfers which can be executed simultaneously. This is an extremely powerful technique for improving performance, allowing data and/or results to be stored while calculating the next set. All these instructions execute in a single cycle and the destination is either of the 56 bit accumulators.

ADDL	Shift Left and add Accumulators
ADDR	Shift Right and Add Accumulators
AND	Logical AND
ANDI	AND Immediate with Control Register*
ASL	Arithmetic Shift Accumulator Left
ASR	Arithmetic Shift Accumulator Right
CLR	Clear Accumulator
CMP	Compare
CMPM	Compare Magnitude
DIV	Divide Iteration
EOR	Logical Exclusive OR
LSL	Logical Shift Left
LSR	Logical Shift Right
MAC	Signed Multiply-Accumulate
MACR	Signed Multiply-Accumulate and Round
MPY	Signed Multiply
MPYR	Signed Multiply and Round
NEG	Negate Accumulator
NORM	Normalize Accumulator Iteration*
NOT	Logical Complement
OR	Logical Inclusive OR
ORI	OR Immediate with Control Register*
RND	Round Accumulator
ROL	Rotate Left
ROR	Rotate Right

*These instructions do not allow parallel data moves.

The DSP56000 arithmetic and logical instructions

The CPMM instruction is useful in determining minima and maxima by setting the condition code bits according to the results of the subtraction of two operands. The NORM instruction performs a single bit normalisation on the contents of an accumulator and updates a specified address register accordingly. This, and the RND instruction for convergent rounding, are extensively used within floating point and scaling operations.

The most powerful operation within the group is the MACR instruction. It performs a signed multiply and accumulate with convergent rounding with two parallel data moves. These four activities are executed within a single cycle.

Bit manipulation

These instructions are familiar to microcontroller programmers and can be split into two groups. They both test the state of any single bit in a memory location, but one group optionally sets, clears or inverts the bit as needed, while the other transfers program flow via a jump or jump to subroutine. Due to their use of the memory buses, parallel data moves are not allowed.

BCLR	Bit Test and Clear
BSET	Bit Test and Set
BCHG	Bit Test and Change
BTST	Bit Test on Memory
JCLR	Jump if Bit Clear
JSET	Jump if Bit Set
JSCLR	Jump to Subroutine if Bit Clear
JSSET	Jump to Subroutine if Bit Set

The DSP56000 bit manipulation instructions

Loop control

There are two loop instructions, DO and ENDDO. Other processor families, like the M68000 and M88000, construct program loops by setting up a loop counter, labeling the start of the loop, decrementing or incrementing the loop counter at the end of the loop and, finally, performing a comparison to decide whether to repeat the loop or drop through to the next instruction. While some support has been provided (e.g. the M68000 DBcc decrement and branch on condition instruction) it still requires a few instructions and their overhead. With the periodic and regular processing to prevent data samples being missed, such overhead cannot be tolerated. To solve this problem, many time critical loops were unfolded and written as straight line sequential code. The disadvantage is the code expansion. A 10 instruction loop repeated 100 times would be replaced by 1,000 instructions: a 100:1 expansion!

The DSP56000 has hardware loop counters which are controlled by the DO and ENDDO instructions. This allows program loops to be set up with no software overhead during the loop. The DO instruction sets up the loop. It saves the LA and LC registers on the stack, so that program loops can be nested, sets up the LA and LC registers accordingly and places the first instruction of the loop on the stack to remove any delay at the loop end. The number of counts can be specified via an indirect address, and allows single instruction loops. DO loops are interruptable.

The ENDDO instruction terminates a DO loop prematurely and cleans up the stack as needed.

None of these instructions allow parallel moves, although this restriction does not apply to the instructions that form the loop itself.

MOVE commands

There are five 'MOVE' instructions which transfer data across the X, Y and global data buses as well as the program data bus. Their execution does not effect the condition code register except for the L limiting bit when either accumulator A or B is read. The MOVE instruction can be considered as a data ALU no-op but with its two parallel data moves intact.

LUA	Load updated address
MOVE	Move data registers
MOVEC	Move control registers
MOVEM	Move program memory
MOVEP	Move peripheral data

The DSP56000 MOVE instructions

Program control

This group of instructions include the almost universal jump and conditional jumps to other addresses or subroutines, return from subroutine or interrupt commands, etc. However, there are some interesting newcomers.

Jcc	Jump conditionally
JMP	Jump
JScc	Jump to subroutine conditionally
JSR	Jump to subroutine
NOP	No operation
REP	Repeat next instruction
RESET	Reset onchip peripheral devices
RTI	Return from interrupt
RTS	Return from subroutine
STOP	Stop instruction processing*
SWI	Software interrupt
WAIT	Wait for interrupt (low power standby modes)

The DSP56000 program control instructions

The REP instruction simply repeats the next instruction without refetching it to maximise throughput. The instruction is not interruptable because of this. A single instruction DO loop can be used instead. There are two power down modes controlled by WAIT and STOP. The WAIT instruction halts internal processing until an interrupt appears to wake it up. The STOP instruction halts all internal processing and gates off the clock.

None of this group of instructions support the parallel data moves.

Instruction format

The first sight of assembler source code for the DSP56000 instils a reaction of sheer terror in those more used to 'normal' assemblers. It looks as if the formatting has gone crazy and put instructions in comment fields! In reality, the assembler format is reflecting the processors´ ability to perform parallel operations as shown.

OPCODE	OPERANDS	X Bus Data	Y Bus Data
MAC	X0,Y0,A	X:(R0)+, X0	Y:(R4)+,Y0

The DSP56000 Instruction format

The opcode and operand fields are self explanatory, however, the next two fields must be considered as part of the single assembler instruction as they specify the two associated data moves.

Using special addressing and loop modes

The requirement for multiple tables lends itself neatly to the adoption of multiple address spaces, allowing parallel fetches. The reasoning is similar to the adoption of Harvard architectures previously discussed. It increases bus bandwidth and reduces execution time. The DSP56000 has taken the approach a step further with a program space and two data spaces.

In addition to the more common addressing modes, the DSP56000 provides modulo addressing to allow circular buffers to be created in memory without the software overhead of incrementing a pointer, testing for out of range, and jumping to the start of the table. These functions are performed by hardware with no extra overhead. This provides an elegant way to implement FIFOs, delay lines, sample buffers or shift registers where a simple pointer is moved instead of the data contents.

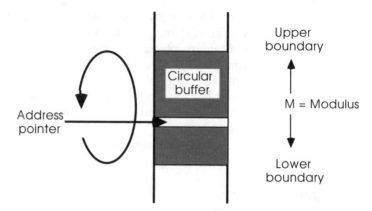

Modulo addressing

In similar fashion, do loops can be performed by dedicated internal hardware instead of the more normal software methods, and again decrease the execution overhead. A hardware loop counter register is provided internally for such applications.

Internal parallelism

The DSP56000 has specific hardware support and parallelism to perform a single cycle 24 bit by 24 bit multiply, a 56 bit addition, two data moves and two address pointer updates in a single cycle of 97 ns.

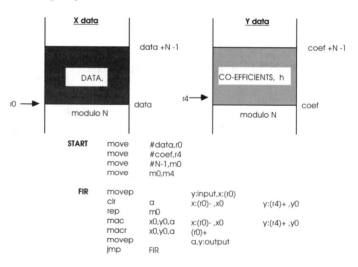

FIR filter program for DSP56000 processor

This feat is another example of the continuing theme of making the most efficient use of the available hardware within a processor. The benefit for DSP56000 users is two instructions executed with no overhead, a reduction in the loop execution time, and an increased data throughput.

The advantage of such facilities can be seen from examination of a typical FIR filter program. It uses just 11 instructions. The essential components of the routine are the 'rep' and 'mac' instructions. The rep instruction is hardware supported and repeats the mac instruction m0 times. The repeated instruction performs the single cycle multiply accumulate with two moves. Without modular addressing and the simultaneous data movements, the loop execution time could be six times greater.

Architectural differences

The DSP56000 is a microprocessor which has been enhanced for DSP applications. Through its microprocessor roots and features, it also makes an exceptionally good embedded controller, however, so does the M68000 and the M88000. The problem is now to understand the other system functions these architectures need, how they perform and their associated trade-offs. These are the next topics for discussion.

DSP96000 — combining integration and performance

The DSP96000 family takes the basic architecture of the DSP56000 and adds extra integration and improved performance to provide an IEEE compatible floating point digital signal processor. While the DSP56000 can cope with the majority of applications with its 24 bit fixed point capabilities, there is a growing need for extra precision and speed which the DSP96000 family offers. The DSP96002 is built using HCMOS technology and has 1,024 words of onchip data RAM split between the X and Y data spaces, 1,024 words of onchip full speed program RAM (PRAM) and two preprogrammed onchip ROMs containing data tables etc. The PRAM areas can be loaded via a bootstrap program from external EPROM. The device has its own onchip support for circuit emulation.

The central processing unit uses three 32 bit execution units, which operate in parallel to perform arithmetic operations, address generation and control and synchronise programs. The device uses a 27 MHz clock, giving a 75 ns cycle time, and can be interfaced with many different memory configurations, depending on the speed needed and the system cost. Static RAMs with access times of 35 ns or better,

DRAM using their special page access methods or video RAM with its serial mode can all be used.

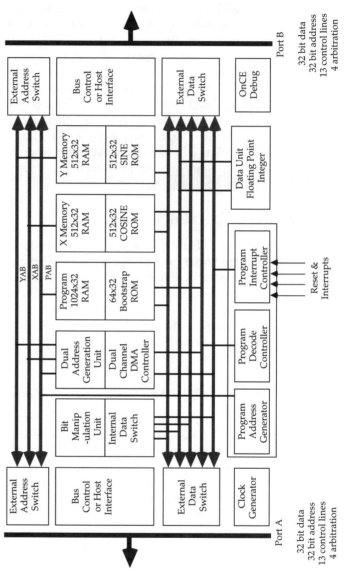

The DSP96002 block diagram

The processing throughput that the device can achieve is 13.5 million instructions per second, where an instruction can involve an instruction prefetch, up to three floating point operations like a 32 bit x 32 bit multiply, two data moves and two address pointer updates. This gives a peak performance of some 108 million operations per second. The IEEE floating point is supported via 10 x 96 bit wide extended precision registers.

Integration has been increased by the addition of two DMA channels, the provision of six onchip memories, two independent expansion bus ports and eight 32 bit buses, three for addresses and five for data. The host interface is supported, as well as the fast interrupt and long interrupt mechanism, as found on the DSP56000. These use 176 of the device´s 214 pins, with 14 reserved for power and 24 for ground.

The instruction set is compatible with that of the DSP56000 but has been expanded to provide floating point and graphics support.

OnCE — a new approach to emulation

The OnCE hardware consists of a serial port and a history buffer, which stores the last five instructions executed. Additional hardware allows single stepping, execution freezing and the setting of breakpoints on any X, Y or P address. The control is located onchip and data is passed to and from a host via the serial line. In effect, the DSP96002 can act as its own emulator in this case, without the problems caused by long cables etc. The system will match any further speed increases because of its integration into the processor itself.

The device can provide a performance upgrade path beyond the simple clock speed increases available to the DSP56000 family and is equally applicable to any system that needs floating point acceleration or extremely fast processing. Applications that can benefit from this include graphics systems, image processing and numeric analysis.

Memory, memory management and caches

This chapter describes the many techniques that are used to build memory subsystems for processors and describes their relevance to the the RISC, CISC and DSP processor families. Each family can use similar techniques to improve its performance, although it does not follow that what is good for a M68000 is automatically good for an M88000.

Achieving processor throughput

Processor throughput depends on three separate activities: opcode and data fetching, instruction execution and finally data storage. For any processor to maintain its maximum performance, this flow of data must be maintained. If there are any wait states in the accesses, performance is lost while the processor is waiting for data. These statements are applicable to any architecture, but the effect of wait states is different depending on the processor architecture.

The graphs overleaf show the effect of wait states on the performance of an MC68000 processor. The CPU performance data is from MacGregor and Rubenstein and has been derived from measurements of bus activity of various systems. It is relative performance, although roughly equivalent to a conservative MIPS rating. While cynics may say that there is no such thing, for this analysis, it is the relative figures that are important. Intuition would indicate that while wait states are detrimental, the processor is more computation bound than bus bound and therefore it is sensible to run the processor faster and tolerate wait states. The reasoning behind this is that CISC processors can take many cycles to execute an instruction and any additional wait states do not greatly contribute to any delays. MacGregor and Rubenstein's analysis showed that on average it took 12.576 clocks per instruction, including an average 2.698 accesses with no wait states. In this instance, the addition of a couple of clocks is not too significant.

The graph shows exactly this. The performance degradation decreases as the number of wait states increases but this is not a linear function. While any removal of wait states improves performance, closer examination confirms that it is beneficial to run the processor faster, although incurring an additional wait state.

Consider the performance of an 8 MHz processor with three wait states: from the graph, it has a performance rating of 0.39. Running a 10 MHz processor with four wait states gives a rating of 0.43, some 16% improvement. If the system is performing a lot of computational work, this value will be even greater.

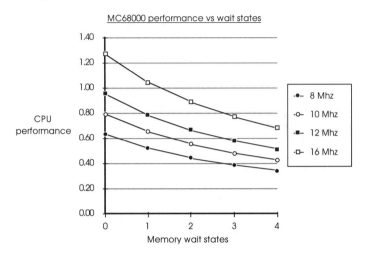

(*based on data from MacGregor and Rubenstein,*
'A performance analysis of MC68020- based systems',
IEEE Micro December 1985)

The effect of wait
states on a CISC
processor

Unfortunately, this may not be so beneficial with RISC processors. RISC architectures simplify the instruction set so that an instruction has to be fetched on every cycle to maintain the data flow into the processor. This effectively means that at 20 MHz, memory is accessed every 50 ns and at the higher clock speeds, such as 40 MHz, this increases to every 25 ns. This places tremendous design constraints on the external memory systems. To maintain the constant flow of instructions into the processor requires a separate bus with a very high bus utilisation (95–100%), hence the Harvard bus architecture with separate data and instruction buses. The insertion of a single wait state halves system performance, two wait states reduce it by two thirds and so on. Examination shows that a RISC architecture, or any processor executing single cycle instructions, suffers far more with wait states than a CISC processor. The graph shows that it is better to run a processor slightly slower but with less wait states. This is the opposite of the case for CISC.

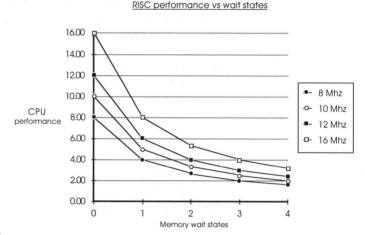

The effect of wait states on the RISC performance

Another useful maxim can be drawn from these graphs: as the average clock per instruction decreases, the more dramatic is the effect that wait states have on performance. RISC systems may have simpler processors but their memory systems are more sophisticated.

DSP processors fall halfway between the two camps. With the DSP56000, its time critical code and data is usually held onchip in wait state free memory. It is only when extra data or code is needed that there is a potential conflict. Depending on the CPI (clock per instruction), the effect can be similar to either CISC or RISC.

Partitioning the system

One way of making the most of the memory resources available is to partition the memory map so that the most appropriate memory is used to maximise performance. The idea is to have different types of memory with different access speeds depending on the performance versus cost trade–offs. These are applicable across the board.

Analysis of systems shows that instructions are fetched more often that data, usually irrespective of the architecture. It follows that it is more beneficial to have faster memory allocated to instructions than data. With a CISC processor that shares a single bus between data and instruction fetches, any delay degrades performance. MacGregor and Rubenstein established that for the M68000 family, there were twice as many op code fetches as data accesses and, therefore, a wait state on any data access was only equivalent to 0.33 of a total system wait state.

With multiple bus architectures like the M88000 and DSP56000, this effect is enhanced. The MC88100 Harvard architecture decouples I/O and data fetches from the instruction traffic and allows both streams to access memory independently. Its separate data unit is pipelined and therefore three to four data fetches can be outstanding before the processor is stalled. A wait state on data fetches may not impact performance at all, provided data accesses are not performed consecutively. With the load-store architecture, this is usually the case. The DSP56000 family, with its multiple buses, offers even more scope for such manipulation.

Shadow RAM

Shadow RAM is a simple technique which is applicable to any processor architecture where essential software or data is stored in non volatile memory, but with the penalty of slow access times. With an operating system kernel stored in an EPROM (erasable programmable read only memory), every system call invariably results in one or more calls to the system routines stored in EPROM. Any removal of wait states from these accesses is extremely beneficial to performance. By copying EPROM contents into RAM on system initialisation, the processor can access the faster RAM and retrieve its lost performance. This overlapping memory bank is called shadow RAM. This technique is used with microcontrollers where EPROM contents are automatically transferred to internal memory on power up, as well as CISC, RISC and DSP processors. The DSP56001 is a version of the DSP56000 that has RAM instead of ROM, and it uses the same technique to load its RAM on power up.

The MC68020 and MC68030 can use shadow RAM, but also use it with dynamic bus sizing so that only a single byte wide EPROM is used to source the information for a 32 bit wide shadow RAM. Only one EPROM is needed, reducing system cost. The processor accesses the boot routines from the EPROM, aibeit slower due to the multiple bus cycles needed, copies the contents into the shadow RAM, and then jumps to the software.

All these techniques map the shadow RAM into a different location. This either needs software that is relocatable or specially written for shadow RAM addressing. For some systems, like the IBM PC, this is not possible due to compatibility restrictions. Here, the shadow RAM must occupy the same address as the original EPROM. This requires special hardware to implement it, like the Motorola MCS2120, PC-AT memory controller. It implements shadow RAM with two independent 128 Kbyte segments — one for the normal BIOS and a second for BIOS extensions. Each segment is enabled before the BIOS code is transferred via a series of consecutive reads and writes at the same address. The MCS2120 routes the cycles accordingly — reads go to the EPROM and writes to the shadow RAM. After the transfer is completed, writing into a configuration register disables the EPROM access and switch the RAM bank into its address space.

DRAM versus SRAM

While most system designers would prefer to implement designs with static RAM, the increased cost of lower memory densities often prevents this. Dynamic memories are cheaper, have higher densities, but require refreshing regularly. The difference between the two memory types is due to the logic used to store each bit of data in a cell. With static memory, five transistors are used to create a 'flip-flop' logic circuit that can be flipped into a low or high state as needed. No refresh is needed and the data remains held as long as power is applied. Dynamic RAM reduces the transistor count to one and uses a small capacitor to store the bit. This increases the memory density, but the storage only remains for a few milliseconds before the capacitor charge disappears through internal leakage, destroying the cell data. This is why DRAM requires regular refreshing. The cell is refreshed whenever the cell is accessed.

There are generally two ways of providing this refresh: the first is to generate a regular processor interrupt and get it to systematically access the memory. This is a tremendous overhead and hinders real–time response, although it does simplify the hardware design. The second is to use a hardware refresh where sections of the memory are accessed periodically by hardware. Typically 4–5% of the memory bandwidth is consumed using these techniques.

The cell transistor count gives a good indicator of the current densities available between memory technologies. The transistor count defines the cell area and hence the densities. Static RAM is usually a factor of 4 smaller in density when compared with the single cell memories like DRAM, EPROM, EEPROM (electrically erasable programmable read only memory) and flash EPROM.

Flash EPROM is a memory technology which can perform bulk erasing quickly, allowing the device to perform as a slow RAM device. It has been hailed as the natural solid state successor to magnetic storage devices, as it can retain data when powered down. What is uncertain is the number of read/write operations that it can sustain before the cells fail. This is extremely important if the technology is going to challenge hard disk drives.

Optimising the DRAM interface

The basic DRAM interface takes the processor generated address, places the high-order bits on to the memory address bus to form the row address and asserts the RAS* signal. This partial address is latched internally by the DRAM. The remaining low-order bits, forming the column address, are then driven onto the bus and the CAS* signal asserted. After the access time has expired, the data appears on the D_{out} pin and is latched by the processor. The RAS* and CAS* signals are then negated. This cycle is repeated for every access.

The majority of DRAM specifications define minimum pulse widths for the RAS* and CAS* and these often form the major part in defining the memory access time.

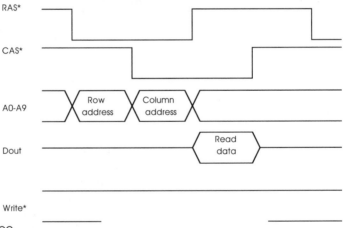

Basic DRAM interface timings

This direct access method limits wait state free operation to the lower processor speeds. DRAM with 100 ns access time would only allow a 12.5 MHz processor to run with zero wait states. To achieve 20 MHz operation needs 40 ns DRAM, which is unavailable today, or fast static RAM which is, at a price.

Page mode operation

One way of reducing the effective access time is to remove the RAS* pulse width every time the DRAM was accessed. It needs to be pulsed on the first access, but subsequent accesses to the same page (i.e. with the same row address) do not require it and so are accessed faster. This is how the 'page mode' versions of most 256 Kbit, 1 Mbit and 4 Mbit memory work. In page mode, the row address is supplied as normal, but the RAS* signal is left asserted. This selects an internal page of memory within the DRAM where any bit of data can be accessed by simply placing the column address and asserting CAS*. With 256 Kbit memory, this gives a page of 1 Kbytes (512 column bits per DRAM row times 16 DRAMs in the array). A 2 Kbyte page is available from 1 Mbit DRAM and 4 Kbyte page with 4 Mbit DRAM.

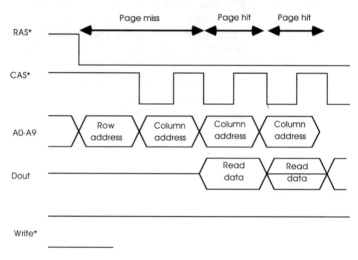

Page mode access

This allows fast processors to work with slower memory and yet achieve almost wait state free operation. The first access is slower and causes wait states, but subsequent accesses within the selected page are quicker with no wait states.

There is one restriction, however. The maximum time that the RAS* signal can be asserted during page mode operation is often specified at about 10 microseconds. In many designs, the refresh interval is frequently adjusted to match this time, so that a refresh cycle will always occur and prevent a specification violation.

Page interleaving

Using a page mode design only provides greater performance when the memory cycles exhibit some form of locality (i.e. stay within the page boundary). Every access outside the boundary causes a page miss and two to three wait states. The secret, as with caches, is to increase the hits and reduce the misses. Fortunately, most accesses are sequential or localised, as found in program subroutines and some data structures. However, if a program is frequently accessing data, the memory activity often follows a code-data-code-data access pattern. If the code areas and data areas are in different pages, any benefit that page mode could offer is lost. Each access changes the page selection, incurring wait states. The solution is to increase the number of pages that are available. If the memory is divided into several banks, each bank can offer a selected page, increasing the number of pages and, ultimately, the number of hits and performance. Again, extensive hardware support is needed.

Typically, a page mode system implements a one, two, or four way system, depending how much memory is in the system. With a four way system, there are four memory banks, each with their own RAS* and CAS* lines. With four Mbytes of DRAM per bank, this would offer 16 Mbytes of system RAM. The 4 way system allows 4 pages to be selected within page mode at any one time. Page 0 is in Bank 1, page 1 in bank 2 and so on, with the sequence restarting after 4 banks, as shown.

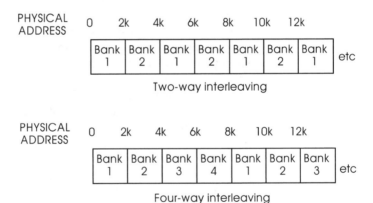

With interleaving and fast page mode devices, inexpensive 85 nsDRAM can be used with a 16 MHz processor to achieve a sub 1 wait state system. With no page mode interleaving, this system inserts two wait states on every access. With the promise of faster DRAM, future designs will be able to offer 20–25 MHz systems with very good performance, without the need for cache memory with its associated costs and complexity.

These pagemode and interleaving techniques are also applicable to SRAM designs and frequently compliment burst fill caches as on the MC68030, MC68040 and PowerPC processors.

Alternative memory systems

However, other memory systems which offer lower cost solutions can be used. Video DRAMs have an onchip shift register which can hold a complete column of data and after the initial slow access can provide the rest of the data in a burst mode until the shift register is empty. This provides data on a single cycle basis once the initial access has been performed. The disadvantage is that every time there is a flow change in the code, a slow access has to be performed to reload the shift register. This slow access may insert four to five waitstates and therefore the average time per access may drop to two clocks per fetch. In addition, video DRAMs are more expensive and have smaller densities than standard DRAMs, which restricts their use to applications requiring small amounts of memory.

Burst mode SRAM

The adoption of burst interfaces by virtually all of today's high performance processors has led to the development of special FSRAMs which include special address generation and data latches to help the designer. Burst interfaces take advantage of page and nibble mode memories which supply data on the first access in the normal time, but can supply subsequently accessed data far quicker. The first access may take two processor clocks but remaining accesses can be made in a single cycle. There are some restrictions to this: the subsequent accesses must be in the same memory page and the processor must have somewhere to store the extra data that can be collected. The obvious solution is to use this burst interface to fill a cache line. The addresses will be in the same page and by storing the extra data in a cache allows a processor to use it at a later date without consuming additional bus bandwidth. The main problem faced by designers with these interfaces is the generation of the new addresses. In most designs the processor will only issue the first address and will hold this constant during the extra accesses. It is up to the interface logic to take this address and increment it with every access. With processors like the MC68030, this function is a straight incremental count. With the MC68040, a wrap-around burst is used where the required data is fetched first and the rest of the line fetched, wrapping around to the line beginning if necessary. Although more efficient for the processor, the wrap-around logic is more complicated.

The solution is to add this logic along with latches and registers to a memory to create a specific part that supports certain bus protocols. The first two members of Motorola's

Protocol specific products are the MCM62940 and MCM62486
32K x 9 fast static RAMs. They are, as their part numbering
suggests, designed to support the MC68040 and the Intel
80486 bus burst protocols. These parts offer access times of 15
and 20 ns.

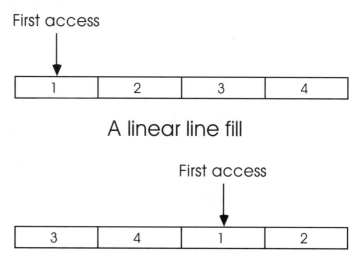

*Linear and wrap-
around line fills*

The MCM62940 has an onchip burst counter that
exactly matches the MC68040 wrap-around burst sequence.
The address and other control data can be stored either by
using the asynchronous or synchronous signals from the
MC68040 depending on the design and its needs. A late write
abort is supported which is useful in cache designs where
cache writes can be aborted later in the cycle than normally
expected, thus giving more time to decide whether the access
should result in a cache hit or be delayed while stale data is
copied back to the main system memory.

The MCM62486 has an onchip burst counter that exactly matches the Intel 80486 burst sequence, again removing
external logic and time delays and allowing the memory to
respond to the processor without the need for the wait state
normally inserted at the cycle start. The chip logic automatically inserts a delay to conform with the processor timing
without necessarily inserting a wait state, thus improving
performance. In addition, it can switch from read to write
mode while maintaining the address and count if a cache read
miss occurs, allowing cache updating without restarting the
whole cycle.

Big vs. little endian organization

Not all processors store data in the same way and this can cause problems when data is transferred between these processors. There are two schemes known as little and big endian which are commonly used. Intel and DEC use the little endian method while Motorola and IBM use the big endian organization. The big and little endian names come from Swift's satire*Gulliver's travels* where a nation was split into two groups depending on whether they ate their eggs starting at the little or big end.

The two methods differ in the location of the most significant and least significant bytes within a word. With a little endian organization, the first byte is the least significant one and the last byte is the most significant one. This is reversed with the big endian organization.

The different organizations can cause problems when data is transferred from a little to a big endian environment and vice versa. Data is reversed and if not correctly re-organized, will be wrongly interpreted.

This problem can even become more annoying when the effect of bit testing is examined. A C program can test bit 15 of a32 bit word but will get different results with a little and big endian environments because the contents of bit 15 are different. In an effort to ease these problems and reduce the overhead of re-organizing every data transfer, many processors offer the ability to support either organization.

Multiplexing addresses

The PowerPC architecture uses primarily a big endian byte order, i.e. an address points to the most significant byte of a value in memory. This can cause problems with other processors that use the alternative little endian organization, where an address points to the least significant byte. The MPC601 solves this problem by providing a mode switch which causes all data memory references to be performed in little-endian fashion.

This is done by swapping address bit lines instead of using data multiplexers. As a result, the byte swapping is not quite what may be expected and varies depending on the size of the data. It is important to remember that swapping the address bits only re-orders the bytes and not the individual bits within the bytes. The bit order remains constant.

The diagram shows the different storage formats for big and little endian double words, words, half words and bytes. The most significant byte in each pair is shaded to highlight its position. Note that there is no difference when storing individual bytes.

As the byte swapping is only concerned with data storage and does not change the instruction format, it is possible to change the endian organization dynamically, although care must be exercised. It is prudent to synchronise the processor with a 'sync' instruction to ensure that all outstanding data accesses have completed before switching the organization. The recommended sequence consists of three sync instructions before and after the control bit is changed. Failure to do this could cause data corruption. A big endian data format may become a little endian format, or vice versa, and be wrongly interpreted.

An alternative solution for processors that do not implement the mode swapping, such as the MPC603, is to use the load and store instructions that byte reverse the data as it moves from the processor to the memory and vice versa.

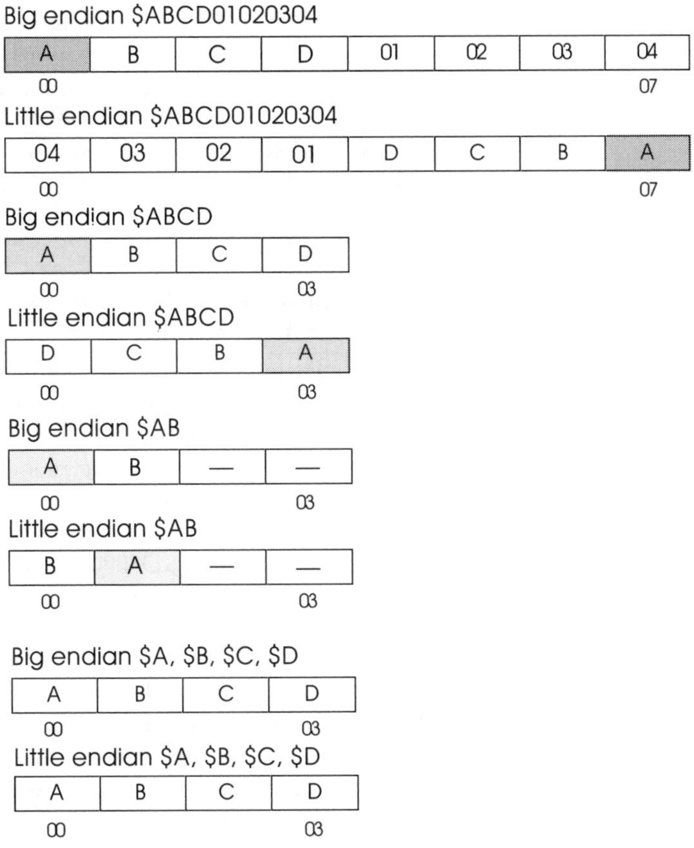

Big endian $ABCD01020304

A	B	C	D	01	02	03	04
00							07

Little endian $ABCD01020304

04	03	02	01	D	C	B	A
00							07

Big endian $ABCD

A	B	C	D
00			03

Little endian $ABCD

D	C	B	A
00			03

Big endian $AB

A	B	—	—
00			03

Little endian $AB

B	A	—	—
00			03

Big endian $A, $B, $C, $D

A	B	C	D
00			03

Little endian $A, $B, $C, $D

A	B	C	D
00			03

Big versus little endian
memory organization

Memory management

Hardware memory management is usually implemented by a MMU(memory management unit) to meet at least one of four system requirements:

1.   The need to extend the current addressing range.

The often perceived need for memory management is usually the result of prior experience or background, and centres on extending the current linear addressing range. The Intel 80x86 architecture is based around a 64 Kbyte linear addressing segment which, while providing 8 bit compatibility, does require memory management to provide the higher order address bits necessary to extend the processor´s address space. Software must track accesses that go beyond this segment, and change the address accordingly. The M68000 family has at least a 16 Mbyte addressing range and does not have this restriction. The M88000 family has an even larger 4 Gbyte range. The DSP56000 has a 128 Kword (1 word = 24 bits) address space, which is sufficient for most present day applications, however, the intermittent delays that occur in servicing an MMU can easily destroy the accuracy of the algorithms. For this reason, the linear addressing range may increase, but it is unlikely that paged or segmented addressing will appear in DSP applications.

2.   To remove the need to write relocatable or position-independent software.

Many systems have multitasking operating systems where the software environment consists of modular blocks of code running under the control of an operating system. There are three ways of allocating memory to these blocks. The first simply distributes blocks in a pre defined way, i.e. task A is given the memory block from $A0000 to $A8000, task B is given from $C0000 to $ D8000, etc. With these addresses, the programmer can write the code to use this memory. This is fine, providing the distribution does not change and there is sufficient design discipline to adhere to the plan. However, it does make all the code hardware and position dependent. If another system has a slightly different memory configuration, the code will not run correctly.

To overcome this problem, software can be written in such a way that it is either relocatable or position independent. These two terms are often interchanged but there is a difference: both can execute anywhere in the memory map, but relocatable code must maintain the same address offsets between its data and code segments. The main technique is to avoid the use of absolute addressing modes, replacing them with relative addressing modes. The M68000, M88000 and DSP56000 families all have this basic architectural support.

If this support is missing or the compiler technology cannot use it, memory management must be used to translate the logical program addresses and map them into physical memory. This effectively realigns the memory so that the processor and software think that the memory is organized specially for them, but in reality is totally different.

3.　　**To partition the system to protect it from other tasks, users, etc.**

To provide stability within a multitasking or multiuser system, it is advisable to partition the memory so that errors within one task do not corrupt others. On a more general level, operating system resources may need separating from applications. Both the M68000 and M88000 families can provide this partitioning through the use of the function codes or by the combination of the user/supervisor signals and Harvard architecture. This partitioning is very coarse, but is often all that is necessary in many cases. For finer grain protection, memory management can be used to add extra description bits to an address to declare its status. If a task attempts to access memory that has not been allocated to it, or its status does not match (e.g. writing to a read only declared memory location), the MMU can detect it and raise an error to the supervisor level.

4.　　**To allow programs to access more memory than is physically present in the system.**

With the large linear addressing offered by today´s 32 bit microprocessors, it is relatively easy to create large software applications which consume vast quantities of memory. While it may be feasible to install 64 Mbytes of RAM in a workstation, the costs are expensive compared with a 64 Mbyte winchester disk. As the memory needs go up, this differential increases. A solution is to use the disk storage as the main storage medium, divide the stored program into small blocks and keep only the blocks in processor system memory that are needed.

As the program executes, the MMU can track how the program uses the blocks, and swap them to and from the disk as needed. If a block is not present in memory, this causes a page fault and forces some exception processing which performs the swapping operation. In this way, the system appears to have large amounts of system RAM when, in reality, it does not. This virtual memory technique is frequently used in workstations and in the UNIX operating system.

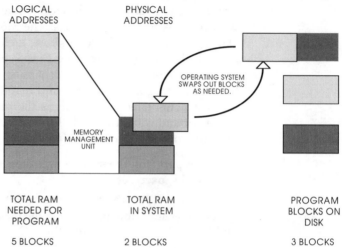

LOGICAL
ADDRESSES

PHYSICAL
ADDRESSES

OPERATING SYSTEM
SWAPS OUT BLOCKS
AS NEEDED.

MEMORY
MANAGEMENT
UNIT

TOTAL RAM NEEDED FOR PROGRAM	TOTAL RAM IN SYSTEM	PROGRAM BLOCKS ON DISK
5 BLOCKS	2 BLOCKS	3 BLOCKS

*Using virtual memory
to support large
applications*

Disadvantages of memory management

Given that memory management is necessary and beneficial, what are the trade-offs ?

The most obvious is the delay it inserts into the memory access cycle. Before a translation can take place, the logical address from the processor must appear. The translation usually involves some form of table look up, where the contents of a segment register or the higher order address bits are used to locate a descriptor within a memory block. This descriptor provides the physical address bits and any partitioning information such as read only etc. These signals are combined with the original lower order address bits to form the physical memory address. This look up takes time, which must be inserted into the memory cycle, and usually causes at least one wait state. This slows the processor and system performance down.

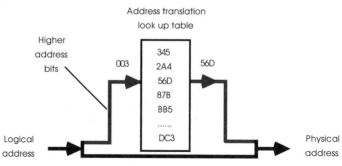

Address translation
look up table

Higher
address
bits

003   56D

| 345 |
| 2A4 |
| 56D |
| 87B |
| BB5 |
| ...... |
| DC3 |

Logical
address

Physical
address

*Address translation
mechanism*

In addition, there can be considerable overheads in managing all the look up tables and checking access rights etc. These overheads appear on loading a task, during any memory allocation and when any virtual memory system needs to swap memory blocks out to disk. The required software support is usually performed by an operating system. In the latter case, if the system memory is very small compared with the virtual memory size and application, the memory management driver will consume a lot of processing and time in simply moving data to and from the disk. In extreme cases, this overhead starts to dominate the system which is working hard but achieving very little. The addition of more memory relieves the need to swap and returns more of the system throughput to executing the application.

Segmentation and paging

There are two methods of splitting the system memory into smaller blocks for memory management. The size of these blocks is quite critical within the design. Each block requires a translation descriptor and therefore the size of the block is important. If the granularity is too small (i.e. the blocks are 1–2 Kbytes), the number of descriptors needed for a 4 Gbyte system is extremely large. If the blocks are too big, the number of descriptors reduces but granularity increases. If a program just needs a few bytes, a complete block will have to be allocated to it and this wastes the unused memory. Between these two extremes lies the ideal trade-off.

A segmented memory management scheme has a small number of descriptors but solves the granularity problem by allowing the segments to be of a variable size in a block of contiguous memory. Each segment descriptor is fairly complex and the hardware has to be able to cope with different address translation widths. The memory usage is greatly improved, although the task of assigning memory segments in the most efficient way is difficult.

This problem occurs when a system has been operating for some time and the segment distribution is now right across the memory map. The free memory has been fragmented into small chunks albeit in large numbers. In such cases, the total system memory may be more than sufficient to allocate another segment, but the memory is non-contiguous and therefore not available. There is nothing more frustrating, when using such systems, as the combination of '2 Mbytes RAM free' and 'Insufficient memory to load' messages when trying to execute a simple utility. In such cases, the current tasks must be stopped, saved and restarted to repack them and free up the memory. This problem can also be found with file storage systems which need contiguous disk sectors and tracks.

A paged memory system splits memory needs into multiple, same sized blocks called pages. These are usually 1–2 Kbytes in size, which allows them to take easy advantage of fragmented memory. However, each page needs a descriptor, which greatly increases the size of the look up tables. With a 4 Gbyte logical address space and 1 Kbyte page size, the number of descriptors needed is over 4 million. Each descriptor would be 32 bits (22 bits translation address, 10 bits for protection and status) in size and the corresponding table would occupy 16 Mbytes! This is a little impractical, to say the least. To decrease the amount of storage needed for the page tables, multi-level tree structures are used. Such mechanisms have been implemented in the MC68851 paged memory management unit (PMMU), the MC68030 processor and the MC88200 CMMU.

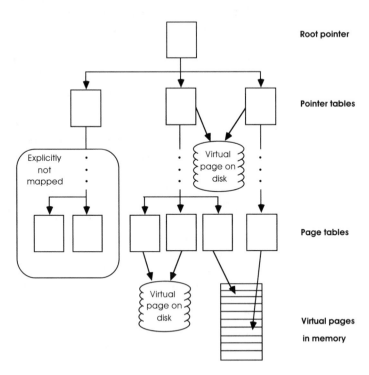

Using trees for descriptor tables

Trees work by using a dividing the logical address into fields and using each of the fields to successively reference into tables until the translation address is located. This is then concatenated with the lower order page address bits to complete a full physical address. The root pointer forms the start of the tree and there may be separate pointers for user and supervisor use.

The root pointer points to separate pointer tables, which in turn point to other tables and so on, until the descriptor is finally reached. Each pointer table contains the address of the next location. Most systems differ in the number of levels and the page sizes that can be implemented. Bits can often be set to terminate the table walk when large memory areas do not need to be uniquely defined at a lower or page level. The less levels, the more efficient the table walking.

The next diagram shows the three-level tree used by the MC88200 CMMU. The logical 32 bit address is extended by a user/supervisor bit which is used to select one of two possible segment table bases from the user and supervisor area registers. The segment number is derived from bits 31 to 22 and is concatenated with a 20 bit value from the segment table base and a binary '00' to create a 32 bit address for the page number table. The page table base is derived similarly until the complete translation is obtained. The remaining 12 bits of the descriptors are used to define the page, segment or area status in terms of access and more recently, cache coherency mechanisms. If the attempted access does not match with these bits (e.g. write to a write protected page), an error will be sent back to the processor.

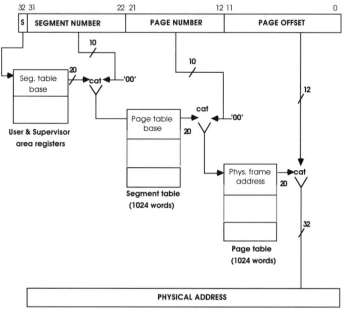

The MC88200 table walking mechanism

Area		WT	0	G	CI	0	0	U	WP	V
	Sup'v. Segment Table	WT	0	G	CI	0	0	U	WP	V
	User Segment Table Base	WT	0	G	CI	0	0	U	WP	V

Segment		WT	SP	G	CI	0	0	U	WP	V
	Page Table Base	WT	SP	G	CI	0	0	U	WP	V

Page		WT	SP	G	CI	0	M	U	WP	V
	Page Frame Base	WT	SP	G	CI	0	M	U	WP	V

V	Valid	CI	Cache inhibit
WP	Write protect enable	G	Global - snoop
U	Used	SP	Supervisor only
M	Modified	WT	Write through

MC88200 memory management descriptors

The next two diagrams show a practical implementation using a two-level tree from an MC68851 PMMU. The example task occupies a 1 Mbyte logical address map consisting of a code, data and stack memory blocks. With a single-level table, this would need 512 descriptors, assuming a 2 Kbyte page size. With the two-level scheme shown, only the needed pages are mapped. All other accesses cause an exception to allocate extra pages as necessary. This type of scheme is often called demand paged memory management.

With the two-level table mechanism, only 44 entries are needed, occupying only 208 bytes, with additional pages increasing this by 4 bytes. The example shows a relatively simple two-level table but up to five levels are often used. This leads to a fairly complex and time consuming table walk to perform the address translation.

To improve performance, a cache or buffer is used to contain the most recently used translations, so that table walks only occur if the entry is not in the address translation cache (ATC) or translation look-aside buffer (TLB). This makes a lot of sense — due to the location of code and data, there will be frequent accesses to the same page and by caching the descriptor, the penalty of a table walk is only paid on the first access. However, there are still some further trade-offs to consider.

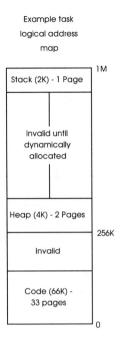

Example task
logical address
map

The required memory map

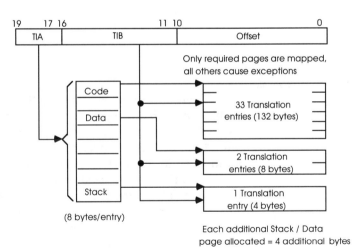

(8 bytes/entry)

Each additional Stack / Data
page allocated = 4 additional bytes

N.B   Multi-level tables require only 44 entries
(total = 208 bytes) to map the 1Mbyte
task space:

Example multilevel tree

Multitasking and user/supervisor conflicts

Using ATCs or TLBs, (for example with the MC68851, MC68030 and MC88200 processors) decreases memory management delays considerably. However, consider the dilemma faced when they are used within a multitasking system as shown.

Two tasks with the same logical address are mapped into different physical locations. Task A executes for the first time: the TLBs are loaded with the correct translations and the system performs correctly. A context switch occurs and task B is loaded with the same logical address. The problem is — how does the MMU know whether its TLB contents are valid?

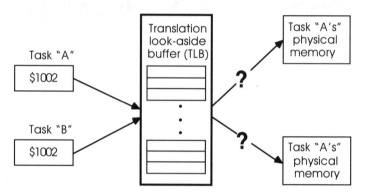

A multitasking memory management dilemma

There are three solutions to this. The first involves flushing the TLB during the context switch so that task B can start with a clean buffer. This is simple but forces several table walks to be performed to load the TLB. The second requires additional information to appended to the each entry, so that a task identity can be used to resolve any conflict. Here, translation entries need not be flushed from the cache and, thus performance is improved. The third involves preloading the TLB during the context switch with the relevant task descriptors so that the tablewalking overhead is removed.

The first solution is simple but has the impact of external table walks. The second needs hardware and effectively splits the TLB into several smaller TLBs and potentially suffers the problem of insufficient TLB entries forcing further table walks. The third has the overhead of adding to the task data that must be saved and reinstated. Which solution is best?

There are no obvious indicators. The MC68851 PMMU implemented the second solution with its 'task aliasing' feature but it was little used (as were about half of the PMMU's facilities) and was not supported on the MC68030 MMU. It is difficult to accurately define the reason, but it was probably due to a combination of silicon real estate limitations on the MC68030 design (its ATC held only 22 entries, compared with 64 for the MC68851), lack of use and the effective reduction of the TLB entries per task. Between the appearance of the MC68851 and the MC68030, programs had grown considerably, processing performance had increased and systems had taken advantage of the new 256 kbit and 1 Mbit DRAM technology to increase the available memory. In such conditions, eight entries per task is simply not big enough. The technique of preloading TLB entries and increasing the number of entries per task would be better. It is interesting to note that operating system page sizes have increased from 1 to 4 Kbytes, with talk of even 8 or 16 Kbytes for the next generation.

A second problem with TLBs and ATCs concerns the updating and overwriting of entries. Consider an operating system with system memory, application memory and I/O areas. All are mapped correctly. During task A's execution time, it accesses many pages and eventually fills up the TLB with all its entries. It then makes a system call to the operating system, the processor switches into supervisor mode and starts to perform the request. With no current TLB entries, this causes an external table walk. Once completed, the request handler can now be accessed and the system call completed. It is extremely likely that multiple table walks would need to be carried out, and these would greatly increase the call execution time. Once completed, task A can restart and load in its TLB entries, overwriting the supervisor entries. The same system call is made and the table walking and subsequent overwriting repeated. This can cause a extremely poor system call response. This scenario can occur when accessing physical I/O such as memory mapped graphics etc.

The MC68030 solves this problem by providing two transparent windows which can be used to directly access two blocks of memory. These can be from 16 Mbytes to 2 Gbytes in size and can be user/supervisor restricted, cache inhibited and write protected, if necessary. They do not use any of the MMU functions and can be dedicated solely to operating system use.

The MC88200 solution is to provide two caches — the paged address translation cache for applications and a separate block address translation cache for supervisor and operating system use. With the multiple CMMU configurations normally used with the MC88100 RISC processor to support its Harvard architecture, there are separate MMUs and ATCs for instructions and data.

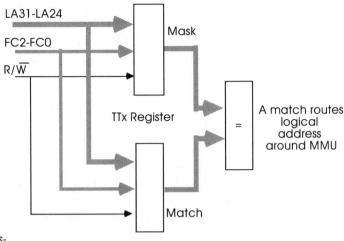

*The MC68030 trans-
parent windows*

Table walking and RISC architectures

Table walking is a good mechanism for loading address descriptors because it reduces the table searching to an
absolute minimum. However, in doing so, the accesses it
performs are pseudo random in nature and the memory
addresses are frequently distributed across a wide range of
the memory map. With RISC processors or indeed any fast
processor that relies on caches, burst fill memory and so on, a
successsion of random memory accesses is probably the most
inefficient way of accessing memory. In an effort to reduce this
loading, one such method is the hashing mechanism used in
the PowerPC architecture.

The hashing mechanism

If there are no matching entries in either the onchip
BAT registers or TLBs, an external PTE search is needed. This
is often referred to as a 'table walk' because it requires searching through sets of tables to find the required data. This
process is often needed to investigate several linked tables
and uses many memory accesses to complete the walk. This
table walk can either be done automatically, as with the
MPC601, or by software, as is the case with the MPC603. In
both cases, the operating system must set up the processor to
help navigate the tables.

Both processors use hashing to generate a pair of PTEG
addresses which should contain the required PTE. This removes the need for an incremental table walk, where one table
is searched for an index which is then added to part of the

logical address to reference the next table, and so on, until the PTE is reached. Hashing can appear to be very complex — but it is very simple in operation.

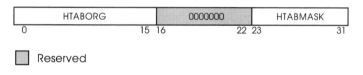

Reserved

The SDR1 register format

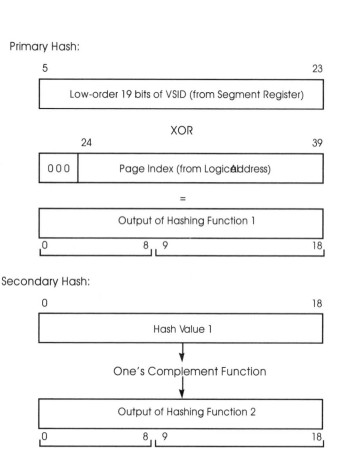

Generating primary and secondary hashing functions

The starting point is the SDR1 register, which has two fields: HTABORG and HTABMASK. HTABORG is the upper 16 bits of the start or origin of the PTEGs. The HTABMASK is a field which determines how many bits from the hashing function are used as the index into the collection of PTEGs located by the address in HTABORG. This combination of the value of HTABORG and HTABMASK determines the number of PTEs that are contained in the collection of PTEGs. While there are minimum recommendations and sizes, the more PTEs, the less likelihood of having no PTEs available to map a VSID.

Two hashing functions are generated, as shown in the diagram. The secondary hash is simply the one's complement of the primary hash. They are automatically generated by hardware within the processor.

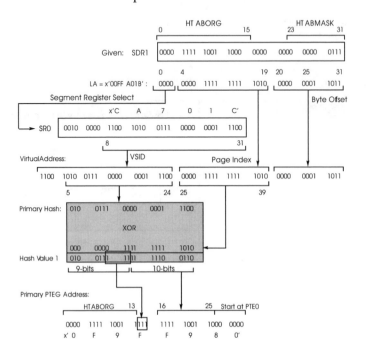

An example or primary hashing

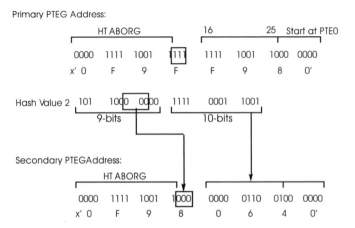

An example of secondary hashing

Given these two hashing values and the contents of SDR1, the MMU uses them as shown to generate two addresses which will point to two individual PTEGs. Essentially, the higher order bits of HTABORG are combined with the hash value which is generated from a logical manipulation of the VSID and page index. These PTEGs are searched in turn, starting with the one pointed to by the results of the primary hash. If this is not successful, the secondary hash is used to access a different PTEG. In both cases, each PTE hash bit is checked to make sure that it complies with the method used to access it. In this way, a PTE can be located by searching only two PTEGs irrespective of the number of PTEGs that have been created.

If searching both the primary and secondary PTEGs does not find a valid PTE, a page fault exception is generated. The operating system then generates a PTE and inserts it into either the primary or secondary PTEG. It has to use the same hashing algorithm to identify the correct PTEG. This is usually implemented in software for the MPC601; the MPC603 has two registers which contain the primary and secondary hashed addresses. If it uses a different algorithm, the PTE will not be located. After setting up the PTE correctly, the access can be restarted, the PTE located, used to update the TLB and perform the address translation.

Advantages of hashing

The hashing mechanism appears to be complicated compared to the more traditional table walking mechanism that is normally used. With this method, the PTE is located by a series of accesses to external memory to a series of linked tables. The difficulty with this approach is that the walking involves several memory accesses to different addresses. After each address is accessed, the contents are concatenated with part of the logical address to form a pointer to the next table level. This table is then accessed and concatenated to create the next pointer and so on, until eventually, the physical address is created.

This type of pseudo-random access is extremely expensive for single cycle processors that use dynamic memory. The problem is that the first access is always the slowest and cannot take advantage of the faster burst mode. As a result, each memory access results in the slowest type of access with no option of using the burst.

The advantage offered by the hashing approach is that although the number of accesses may be increased — there are a maximum of eight PTEs to be checked for each hashed address — the data is sequential and therefore can be loaded into the processor using the burst mode using single cycle access. The overall time compared to a three-level table walk is far less.

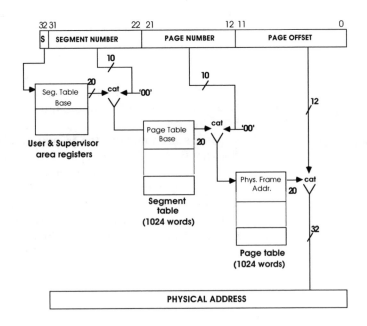

Three level table walking

With a 5-1-1-1 memory system where the first access takes five clocks and the subsequent burst accesses are single cycle, it would take eight clocks to access the first four PTEs using a 64 bit data bus. With eight PTEs per PTEG and assuming equal distribution, the average number of accesses will be four and so the 8 clocks will become the average time to search a hashed PTEG. With a three-level table walk which requires three external memory accesses which cannot take advantage of the burst mechanism, the averag time becomes three times five clocks, giving 15 clocks — almost double that for the hashed mechanism. As the penalty for the first access increases, the improvements offered by hashing improve even more.

This analysis has ignored the case where the second hashed address must be used which would increase the overall average time. However, it should be remembered that its use depends on the amount of space allocated to the page table structures — the more space, the higher the chance that the primary PTEG will hold the needed PTE. This parameter is under software control and is part of the SDR1 register facilities. The supervisor can trade off speed with memory: by allocating more page table space in memory, the hit rate in the primary hashed PTEG increases and the overall average penalty for fetching an external descriptor reduces. With a table walk based MMU, this table walk size is frequently fixed and its search speed is unaffected by the size of the memory allocated to hold page tables.

Instruction continuation versus restart

Another potential problem is thrashing during page swapping, caused by an instruction continuation instead of instruction restart. If, during an instruction execution, a memory access goes across a page boundary and causes a page fault, a page swap is performed. The required page is taken from disk and may replace the old page which contained the original instruction. If the processor uses instruction continuation, there is no problem. The old page is not needed and only the faulting access is attempted again to carry on as before. If the processor has to restart the instruction, the original instruction is needed. This has been swapped out to disk and the subsequent fetch causes another page fault. It is likely that the new page is chosen to swop out because it is the least recently used. The instruction is executed, the data access is attempted causing a page fault etc. The system starts to thrash and may never resolve this situation.

In reality, this can easily be detected by simply enforcing a rule of 'never swapping the previous page' and is therefore never a serious problem. However, if software is ported from a 'continuation' architecture to a 'restart' one, this must be handled.

Memory management and DSP

On first inspection, digital signal processing appears to have no need for memory management at all. The programs are relatively small and do not need virtual memory. They do not run multi-user operating systems and therefore do not need protection. They cannot tolerate the random delays caused by page faults and external table walks which would seriously reduce the signal bandwidth that they can handle, or introduce distortion errors. However, DSP processors are becoming more like general purpose processors and, with the DSP96000, will have 32 bit addressing. It may be that future generations will start to need memory management functions so they can run both DSP and more general applications.

Cache memory

Any comments on cache memory design and its parameters are almost guaranteed to start a debate among engineers, as zealous as any discussion in a synod or other philosophical gathering! This is because there are no wrongs or rights in cache design — it inevitably results in a juggling act involving the resources available. It is therefore important to understand the ideas behind such designs and the trade-off decisions before making instantaneous evaluations and dismissing them. It is always sobering to remember that the simple 256 byte internal cache on the MC68020 gives a performance increase of about 17–25 %!

With faster (>25 MHz) processors, the wait states incurred in DRAM accesses start to dramatically reduce performance. To recover this, many designs implement a cache memory to buffer the processor from such delays. Cache memory systems work because of the cyclical structures within software. Most software structures are loops where pieces of code are repeatedly executed, albeit with different data. Cache memory systems store these loops so that after the loop has been fetched from main memory, it can be obtained from the cache for subsequent executions. Accesses from cache are faster than from main memory and thus increase system throughput. This means that caches cannot be 100% efficient — the first accesses always goes through to main memory.

Cache size and organization

There are several criteria associated with cache design which affect its performance. The most obvious is cache size and organization — the larger the cache, the more entries that are stored and the higher the hit rate. However, as the cache size increases, the return gets smaller and smaller. In practice, the cache costs and complexity place an economic limit on most designs. As the size of programs increase, larger caches are needed to maintain the same hit rate and hence the 'ideal cache size is always twice that available' comment. In reality, it is the combination of size, organization and cost that really determines the size and its efficiency.

Consider a basic cache operation. The processor generates an address which is fed into the cache memory system. The cache stores its data in an array with an address tag. Each tag is compared in turn with the incoming address. If they do not match, the next tag is compared. If they do match, a cache hit occurs, the corresponding data within the array is passed on the data bus to the processor and no further comparisons are made. If no match is found (a cache miss), the data is fetched from external memory and a new entry is created in the array. This is simple to implement, needing only a memory array and a single comparator and counter. Unfortunately, the efficiency is not very good due to the serial interrogation of the tags.

A better solution is to have a comparator for each entry, so all entries can be tested simultaneously. This is the organization used in a fully associative cache. In the example, a valid bit is added to each entry in the cache array, so that invalid or unused entries can be easily identified. The system is very efficient from a software perspective — any entry can be used to store data from any address. A software loop using only 20 bytes (10 off 16 bit instructions) but scattered over a 1,024 byte range would run as efficiently as another loop of the same size but occupying consecutive locations.

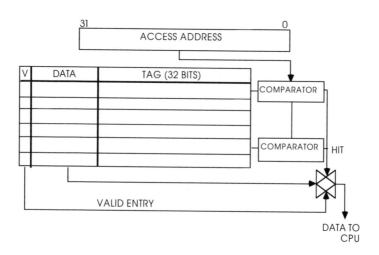

A fully associative cache design

The disadvantage of this approach is the amount of hardware needed to perform the comparisons. This increases in proportion to the cache size and therefore limits these fully associative caches to about 10 entries. Address translation caches are often fully associative.

The fully associative cache locates its data by effectively asking n questions where n is the number of entries within it. An alternative organization is to assume certain facts derived from the data address so that only one location can possibly have the data and only one comparator is needed, irrespective of the cache size. This is the idea behind the direct map cache.

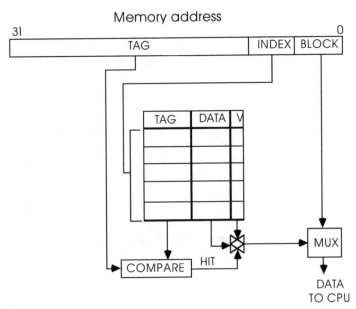

Memory address

Direct mapped cache

The address is presented as normal but part of it is used to index into the tag array. The corresponding tag is compared and, if there is a match, the data is supplied. If there is a cache miss, the data is fetched from external memory as normal and the cache updated as necessary. The example shows how the lower address bits can be used to locate a byte, word or long word within the memory block stored within the array. This organization is simple from the hardware design but can be inefficient from a software viewpoint.

The index mechanism effectively splits external memory space into a series of consecutive memory pages, with each page the same size as the cache. Each page is mapped to resemble the cache and therefore each location in the external memory page can only correspond with its own location in the cache. Data that is offset by the cache size thus occupies the same location within the cache, albeit with different tag values. This can cause bus thrashing.

Consider a case where words A and B are offset by the cache size. Here, every time word A is accessed, word B is discarded from the cache. Every time word B is accessed, word A is lost. The cache starts thrashing and the overall performance is degraded. The MC68020 is a typical direct mapped cache.

A way to solve this is to split the cache so there are two or four possible entries available for use. This increases the comparator count but provides alternative locations and prevents bus thrashing. Such designs are described as '*n* way set associative', where n is the number of possible locations. Values of 2, 4, 8 are quite typical of such designs. The MC88200 CMMU is organized as a four way set associative cache.

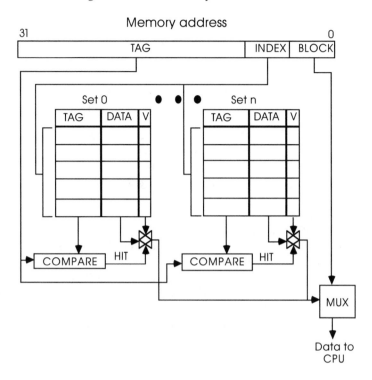

A set associative cache design

The advantage of a set associative cache is its ability to prevent thrashing at the expense of extra hardware. However, all the caches so far described can be improved by further reorganizing so that each tag is associated with a line of long words which can be burst filled using a page memory interface.

The logic behind this idea is based on the sequential nature of instruction execution and data access. Instruction fetches and execution simply involve accesses to sequential memory locations until a program flow change happens due to the execution of a branch or jump instruction. Data accesses often follow this pattern during stack and data structure manipulation.

It follows that if a cache is organized with, say, four long words within each tag line, a hit in the first long word would usually result in hits in the rest of the line, unless a flow change took place. If a cache miss was experienced, it would be beneficial to bring in the whole line, providing this could be achieved in less time than to bring in the four long words individually. This is exactly what happens in a page mode interface. By combining these, a more efficient cache can be designed which even benefits in line code. This is exactly how the MC68030 cache works.

Address bits 2 and 3 select which long word is required from the four stored in the 16 byte wide line. The remaining higher address bits and function codes are the tag which can differentiate between supervisor or user accesses etc. If there is a cache miss, the processor uses its synchronous bus with burst fill to load up the complete line.

	W/O Burst	W/ Burst
Instruction Cache	46%	82%
Data Cache - Reads	60%	72%
Data Cache - R & W	40%	48%

Estimated hit rates for the MC68030 caches

With a complete line updated in the cache, the next three instructions result in a hit, providing there is no preceding flow change. These benefit from being cached, even though it is their first execution. This is a great improvement over previous designs, where the software had to loop before any benefit could be gained. The table above shows the estimated improvements that can be obtained.

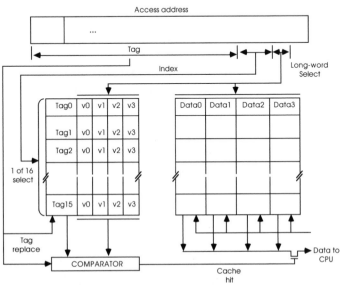

The MC68030 direct mapped burst fill cache

The effect is similar to increasing the cache size. The largest effect being with instruction fetches which show the greatest degree of locality. Data reads are second, but with less impact due to isolated byte and word accesses. Data read and write operations are further reduced, caused by the cache´s write through policy. This forces all read-modify-write and write operations to main memory. In some designs, data accesses are not sequential, in which case, system performance actually degrades when the data cache is enabled — burst filling the next three longwords is simply a waste of time, bus bandwidth and performance. The solution is simple — switch off the cache!

This design is used in most high performance cache designs.

Optimising line length and cache size

This performance degradation is symptomatic of external bus thrashing due to the cache line length and / or burst fill length being wrong and leading to system inefficiencies. It is therefore important to to get these values correct. If the burst fill length is greater than the number of sequential instructions executed before a flow change, data is fetched which will not be used. This consumes valuable external bus bandwidth. If the burst length is greater than the line length, multiple cache lines have to be updated, which might destroy a cache entry for another piece of code that will be executed later. This destroys the efficiency of the cache mechanism and increases the cache flushing, again consuming external bus bandwidth. Both of these contribute to the notorious 'bus thrashing' syndrome where the processor spends vast amounts of time fetching data that it never uses. Some cache schemes allow line lengths of 1, 4, 8, 16 or 32 to be selected, however, most systems use a line and burst fill length of 4. Where there are large blocks of data to be moved, higher values can improve performance within these moves, but this must be offset by any affect on other activities.

Cache size is another variable which can affect performance. Unfortunately, it always seems to be the case that the ideal cache is twice the size of that currently available! The biggest difficulty is that cache size and efficiency are totally software dependant — a configuration that works for one application is not necessarily the optimum for another.

The table shows some efficiency figures quoted by Intel in their 80386 Hardware Reference Manual and from this data, it is apparent that there is no clear cut advantage of one configuration over another.

Common to all the cache organizations is the dependency on fast tag access and comparison. Typical access speeds for tag RAMs are 15–25 ns with 20–25 ns static RAMs forming the data arrays.

Size (K)	Associativity	Line size (bytes)	Hit rate (%)	Performance ratio versus DRAM
1	direct	4	41	0.91
8	direct	4	73	1.25
16	direct	4	81	1.35
32	direct	4	86	1.38
32	2-way	4	87	1.39
32	direct	8	91	1.41
64	direct	4	88	1.39
64	2-way	4	89	1.40
64	4-way	4	89	1.40
64	direct	8	92	1.42
64	2-way	8	93	1.42
128	direct	4	89	1.39
128	2-way	4	89	1.40
12	direct	8	93	1.42

(source: 80386 Hardware Reference Manual)

Cache performance

Logical versus physical caches

Cache memory can be located either side of a memory management unit and use either physical or logical addresses as its tag data. In terms of performance, the location of the cache can dramatically affect system performance. With a logical cache, the tag information refers to the logical addresses currently in use by the executing task. If the task is switched out during a context switch, the cache tags are no longer valid and the cache, together with its often hard-won data must be flushed and cleared. The processor must now go to main memory to fetch the first instructions and wait until the second iteration before any benefit is obtained from the cache. However, cache accesses do not need to go through the MMU and do not suffer from any associated delay.

Physical caches use physical addresses, do not need flushing on a context switch and therefore data is preserved within the cache. The disadvantage is that all accesses must go through the memory management unit, thus incurring delays. Particular care must also be exercised when pages are swapped to and from disk. If the processor does not invalidate any associated cache entries, the cache contents will be different from the main memory contents by virtue of the new page that has been swapped in.

Of the two systems, physical caches are more efficient, providing the cache coherency problem is solved and MMU delays are kept to a minimum. The MC88200 solves the MMU delay by coupling the MMU with the cache system. An MMU translation is performed in conjunction with the cache look up so that the translation delay overlaps the memory access and is reduced to zero. This system combines the speed advantages of a logical cache with the data efficiency of a physical cache.

Most internal caches are now designed to use the physical address (notable exceptions are some implementations of the SPARC architecture which uses logical internal caches).

Unified versus Harvard caches

There is another aspect of cache design that causes great debate among designers and this concerns whether the cache is unified or separate. A unified cache, as used on the Intel 80486DX processors and the Motorola MPC601 PowerPC chip, uses the same cache mechanism to store both data and instructions. The separate or Harvard cache architecture has separate caches for data and instructions. The argument for the unified cache is that its single set of tags and comparators reduces the amount of silicon needed to implement it and thus for a given die area, a larger cache can be provided compared to separate caches. The argument against is that a unified cache usually has only a single port and therefore simultaneous access to both instructions and data will result in one or the other being delayed while the first access is completed. This delay can halt or slow down the processor's ability to execute instructions.

Conversely, the Harvard approach uses more silicon area for the second set of tags and comparators but does allow simultaneousl access. In reality, the overall merits of each approach depend on several factors, and depending where the cross-over points lie, the factors will be in favour of one or other. If software needs to exploit superscalar operation then the Harvard architecture is less likely to impede superscalar execution. If the application has large data and code structures, then a larger unified cache may be better. As with most cache orgnaisation decisions, the only clear way to make a decision is to evaluate using the end application and the test software.

Cache coherency

The biggest challenge with cache design for any system architecture is how to solve the problem of data and cache coherency. This issue arises when data is cached which can be modified by more than one source. The problem is normally only associated with data but can occur with instructions in a page swapping operating system. The stale data situation arises when a copy is held both in cache and main memory. If either copy is modified, the other becomes stale and system coherency is destroyed. The simple solution is to declare shared data as non-cacheable so that any processor access is forced to main memory. This ensures coherency but it couples the processor to slow memory and really defeats the reason for implementing a cache at all. For very simple systems, it is a suitable solution.

There are four typical methods of writing cached data, as shown. However, they all have trade-offs and many systems implement all four on a memory block by block basis to optimise the system performance.

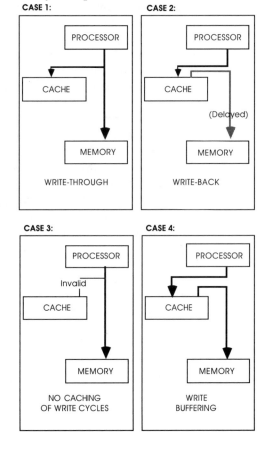

Different write schemes

Case 1: write through

In this case, all data writes go through to main memory and update the system as well as the cache. This is simple to implement but couples the processor unnecessarily to slow memory. If data is modified several times before another master needs it, the write through policy consumes external bus bandwidth supplying data that is not needed. This is not terribly efficient. In its favour, the scheme is very simple to implement, providing there is only a single cache within the system.

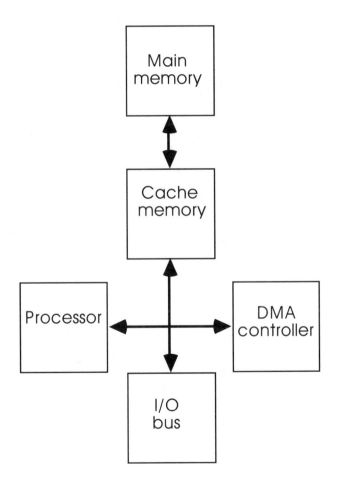

A coherent cache architecture

If there are more than two caches, the stale data problem reappears in another guise. Consider such a system where two processors with caches have a copy of a global variable.

Neither processor accesses main memory when reading the variable, as the data is supplied by the respective caches. Processor A now modifies the variable — its cache is updated, along with the system memory. Unfortunately, processor B´s cache is left with the old stale data, creating a coherency problem. A similar problem can occur within paging systems.

DMA (direct memory access) can modify memory directly without any processor intervention. Consider a UNIX paging system. Page A is in physical memory and is cached. A page fault occurs and page A is swapped out to disk and replaced by page B at the same location. The cached data is now stale. A software solution to this involves flushing the cache when the page fault happens so the previous contents are removed. This can destroy useful cached data and needs operating system support, which can make it non-compatible. The only hardware solution is to force any access to the main memory via the cache, so that the cache can update any modifications.

This provides a transparent solution, but it does force the processor to compete with the DMA channels, and restricts caching to main memory only, with the subsequent reduced performance.

Case 2: write back

In this case, the cache is updated first but the main memory is not updated until later. This is probably the most efficient method of caching, giving 15–20% improvement over a straight write-through cache. This scheme needs a bus snooping mechanism for coherency and this will be described later.

The usual cache implementation involves adding dirty bits to the tag to indicate which cache lines or partial lines hold modified data that has not been written out to main memory. This dirty data must be written out if there is any possibility that the information will be lost. If a cache line is to be replaced as a result of a cache miss and the line contains dirty data, the dirty data must be written out before the new cache line can be accepted. This increases the impact of a cache miss on the system. There can be further complications if memory management page faults occur. However, these aspects must be put into perspective — yes, there will be some system impact if lines must be written out, but this will have less impact on a wider scale. It can double the time to access a cache line, but it has probably saved more performance by removing multiple accesses through to the main memory. The trick is to get the balance in your favour.

Case 3: no caching of write cycles

In this method, the data is written through but the cache is not updated. If the previous data had been cached, that entry is marked invalid and is not used. This forces the processor to access the data from main memory. In isolation, this scheme does seem to be extremely wasteful, however, it often forms the backbone of a bus snooping mechanism.

Case 4: write buffer

This is a variation on the write-through policy. Writes are written out via a buffer to main memory. This enables the processor to update 'main memory' very quickly, allowing it to carry on processing data supplied by the cache. While this is going on, the buffer transfers the data to main memory. The main advantage is the removal of memory delays during the writes. The system still suffers from coherency problems caused through multiple caches.

Another term associated with these techniques is write allocation. A write-allocate cache allocates entries in the cache for any data that is written out. The idea behind this is simple - if data is being transferred to external memory, why not cache it, so that when it is accessed again, it is already waiting in the cache. This is a good idea if the cache is large but it does run the risk of overwriting other entries that may be more useful. This problem is particularly relevant if the processor performs block transfers or memory initialisation. Its main use is within bus snooping mechanisms where a first write-allocate policy can be used to tell other caches that their data is now invalid.

The most important need with these methods and ideas is bus snooping.

Bus snooping

With bus snooping, a memory cache monitors the external bus for any access to data within main memory that it already has. If the cache data is more recent, the cache can either supply it direct or force the other master off the bus, update main memory and start a retry, thus allowing the original master access to valid data. As an alternative to forcing a retry, the cache containing the valid data can act as memory and supply the data directly. As previously discussed, bus snooping is essential for any multimaster system ensure cache coherency.

The bus snooping mechanism used by the MC88100/ MC88200 uses a combination of write policies, cache tag status and bus monitoring to ensure coherency. The next nine diagrams show a typical sequence.

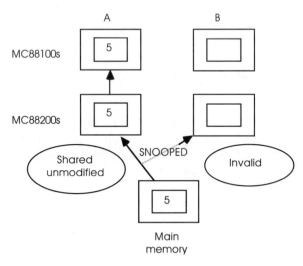

In the first figure, processor A reads data from main memory and this data is cached. Main memory is declared global and is shared by processors A and B. Both these caches have bus snooping enabled for this global memory.

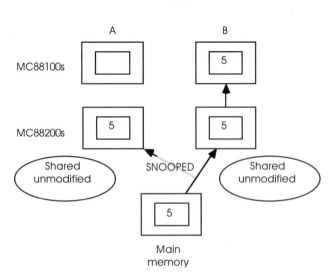

This causes the cached data to be tagged as shared unmodified; i.e. another master may need it and the data is identical to that of main memory. A´s access is snooped by processor B, which does nothing as its cache entry is invalid. It should be noted that snooping does not require any direct processor of software intervention and is entirely automatic.

Processor B accesses main memory, as shown in the next diagram and updates its cache as well. This is snooped by A but the current tag of shared unmodified is still correct and nothing is done.

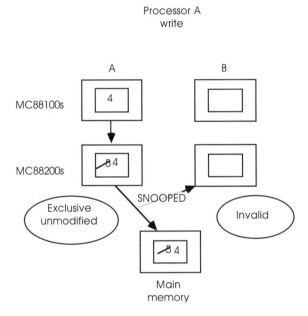

M88000 cache coherency - iii

Processor A then modifies its data as shown in diagram (iii) and by virtue of a first write-allocate policy, writes through to main memory. It changes the tag to exclusive unmodified; i.e. the data is cached exclusively by A and is coherent with main memory. B snoops the access and immediately invalidates its old copy within its cache.

When processor B needs the data, it is accessed from main memory and written into the cache as shared unmodified data. This is snooped by A, which changes its data to the same status. Both processors now know that the data they have is coherent with main memory and is shared.

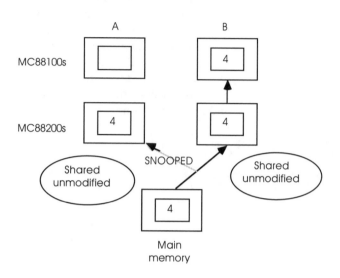

Processor B
read

M88000 cache
coherency - iv

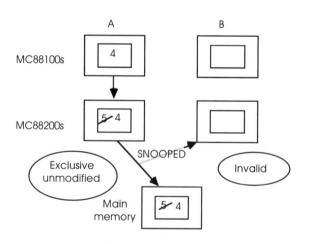

Processor A
write

M88000 cache
coherency - v

Processor A now modifies the data which is written out to main memory and snooped by B which marks its cache entry as invalid. Again, this is a first write-allocate policy in effect.

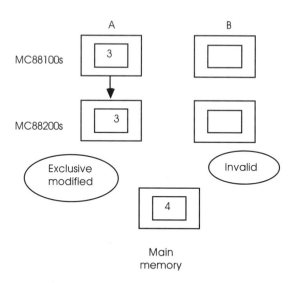

M88000 cache coherency - vi

Processor A modifies the data again but, by virtue of the copyback selection, the data is not written out to main memory. Its cache entry is now tagged as exclusive modified; i.e. this may be the only valid copy within the system.

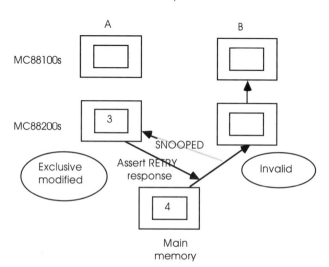

M88000 cache coherency - vii

Processor B tries to get the data and starts an external memory access, as shown. Processor A snoops this access, recognises that it has the valid copy and so asserts a retry response to processor B, which comes off the bus and allows processor A to update main memory and change its cache tag status to shared unmodified.

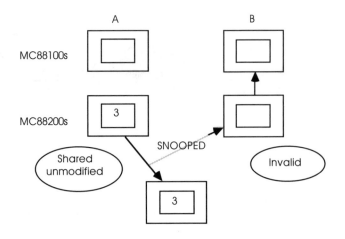

M88000 cache coherency - viii

Once completed, processor B is allowed back on to the bus to complete its original access, this time with the main memory containing the correct data.

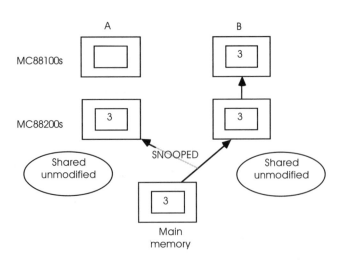

M88000 cache coherency - ix

This sequence is relatively simple, compared with those encountered in real life where page faults, cache flushing, etc. further complicate the state diagrams. The control logic for the CMMU is far more complex than that of the MC88100 processor itself and this demonstrates the complexity involved in ensuring cache coherency within multiprocessor systems.

The problem of maintaining cache coherency has led to the development of two standard mechanisms — MESI and MEI. The MC88100 sequence that has just been discussed is similar to that of the MESI protocol. The MC68040 cache coherency scheme is similar to that of the MEI protocol.

The MESI protocol

The MESI protocol is a formal mechanism for controlling cache coherency using snooping techniques. Its acronym stands for modified, exclusive, shared, invalid and refers to the states that cached data can take. Transition between the states is controlled by memory accesses and bus snooping activity. This information appears on special signal pins during bus transactions.

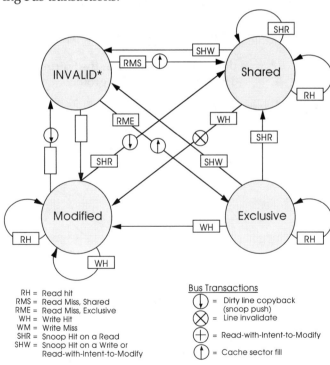

RH = Read hit
RMS = Read Miss, Shared
RME = Read Miss, Exclusive
WH = Write Hit
WM = Write Miss
SHR = Snoop Hit on a Read
SHW = Snoop Hit on a Write or
Read-with-Intent-to-Modify

Bus Transactions

⊕ = Dirty line copyback (snoop push)
⊗ = Line invalidate
⊕ = Read-with-Intent-to-Modify
⊕ = Cache sector fill

* On a cache miss, the old line is invalidated and copied back if modified

MESI cache coherency protocol

The MESI diagram is generic and shows the general operation of the protocol. There are four states that describe the cache contents and its coherence with system memory:

Invalid   The target address is not cached.

Shared   The target address is in the cache and also in at least one other. It is coherent with system memory.

Exclusive   The target address is in the cache but the data is coherent with system memory.

Modified   The target address is in the cache, but the contents has been modified and is not coherent with system memory. No other cache in the system has this data.

The movement from one state is governed by memory actions, cache hits and misses and snooping activity. For example, if a processor needs to write data to a memory address that has a write back policy and cache coherency enabled as part of its page descriptors — controlled by the WIM bits — and causes a cache miss, the processor will move from an invalid state to a modified state by performing a 'read with intent to modify' bus cycle.

The MESI protocol is widely used in multiprocessor designs, for example, in the Futurebus+ interconnection bus. The MPC601 uses this protocol.

The MEI protocol

The MEI protocol — modified, exclusive, invalid — does not implement the shared state and so does not support the MESI shared state where multiple processors can cache shared data.

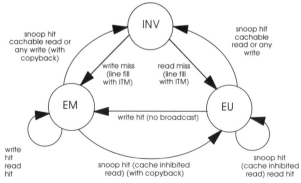

INV - Invalid
EM - Exclusive Modified
EU - Exclusive Unmodified
ITM - Intent-to-modify

MPC603 MEI
coherency diagram

The MPC603 uses this simplified form of protocol to support other intelligent bus masters such as DMA controllers. It is not as good as the MESI bus for true multiprocessor support. On the other hand, it is less complex and easier to implement.

The three states are defined as follows:

Invalid   The target address is not cached.

Exclusive unmodified The target address is in the cache but the data is coherent with system memory.

Exclusive modified   The target address is in the cache, but the contents have been modified and are not coherent with system memory. No other cache in the system has this data.

Note that the cache coherency implementation is processor specific and may change. The two mechanisms described here are the two most commonly used methods for processors and are likely to form the basis of future designs.

One final point: these schemes require information to be passed by the external busses to allow other bus masters to identify the transitions. This requires the hardware design to implement them. If this is not done, these schemes will not work and the software environment may require extensive change and the imposition of constraints. Cache coherency may need to be restricted to cache inhibited or write-through. DMA accesses could only be made to cache inhibited memory regions. The supervisor must take responsibility for these decisions and implementations to ensure correct operation. In other words, do not assume that cache coherency software for one hardware design will work on another. It will, if the bus interface design is the same. It will not if they are different.

Streaming and CWF (critical word first)

When a cache miss occurs, the data must be fetched from memory. However, the memory access is not simply concerned with accessing an individual location—it will load the complete line into the cache. The reasoning for this is simple: if the program requires an instruction, there is a very high probability that it will need the following instructions as well. If this data can be efficiently loaded then, although the processor has suffered a delay in accessing the memory because of the cache miss, it can gain benefit because the next accesses are in the cache. This argument still holds with data, but the probability is reduced.

Line filling also fits in with memory designs which support or implement page mode operation. With this mode, the first access is slower and may take several cycles but subsequent ones can be single cycle. The overall benefit is a reduced access time compared with accessing the individual locations as individual cycles. With a 5-1-1-1 system, it would take only eight clocks to make four consecutive accesses. Performing the same sequence as individual accesses would take 20 clocks — over twice the time!

While this scenario is beneficial to the cache and ulti-mately to the processor, filling cache lines as opposed to an individual access can cause problems. If the required data is at the end of the cache line, the processor will have to wait until the end of the filling sequence to get its data and proceed. In this case, it will wait longer than with a single access. With the previous example, this would be eight clocks compared with five.

The MPC601 and MPC603 both use two techniques to help alleviate these problems. These techniques are known as 'critical word first' and streaming.

Critical word first

This technique changes the order in which the cache line is filled so that the data that the processor requires is fetched first and passed to the processor while it is written into the cache. As a result, the processor is not delayed more than necessary.

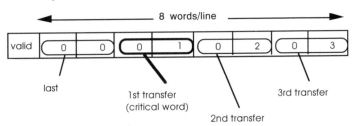

Critical word first
operation

The line is filled by wrapping around, starting from the critical word, continuing to the end of the line and then moving to the start of the line.

Streaming

The critical word first policy is fine for individual accesses but not as efficient as it could be with instruction accesses where the probability is high that the rest of the line will be needed. A technique called streaming is used to alleviate this problem. When the data is read into the cache line, it can also be streamed into the execution units in parallel to help remove delays. The maximum performance with this is obtained when the critical word is the first one in the cache line. As the data is brought in, it can be streamed through and, if this is an instruction stream, processed in parallel with updating the cache.

To fully exploit both of these techniques, it is best to align both data structures and instruction sequences so that they start on a cache line boundary. For data, this may involve padding out the structures to create the correct alignment. For instruction sequences, the technique involves aligning all target and subroutine addresses on the start of a cache line. Some compilers have this support built in while others do not and so require the programmer to manually perform the alignment.

These techniques do have one drawback: they can dramatically increase the memory taken by both programs and data. However, if the goal is to extract the maximum performance, this is the price that must be paid.

Cache control instructions

Cache efficiency could be improved if there was some way that a program could pre-load data and instructions before it was needed. For example, if a program could pre-load the next set of data it needed to process while it was processing the current set, the data access could be overlapped with the processing, greatly reducing the overall execution time.

code	operation	synchronisation
icbi	Instruction cache block invalidate	synchronise instruction stream
isync	Instruction synchronize	synchronise instruction stream
dcbt	Data cache block touch	
dcbtst	Data cache block touch for store	synchronise instruction stream
dcbz	Data cache block set to zero	
dcbst	Data cache block store	
dcbi	Data cache block invalidate	
dcbf	Data cache block flush	

Cache control instructions

The PowerPC architecture provides some cache control instructions which allow a program to invalidate, pre-load or flush a cache line.

Implementing memory systems

While many system designs use cache memory to buffer the fast processor from the slower system memory, it should be remembered that access to system memory is required on the first execution of an instruction or software loop and whenever a cache miss is incurred. If this access is too slow, these overheads greatly diminish the efficiency of the cache and, ultimately, the processor´s performance. Great store is made of cache implementations, but unless the main system memory is fast, system and software performance will degrade. Caches help to regain the performance lost through system memory wait states but they are never 100% efficient. A system with no wait states always provides the best performance. Given that a cache is required, there are effectively two methods of practical implementation — discrete and integrated.

Secondary or level 2 caches

Even with internal caches, many processors still require external caches to provide the best possible external memory bandwidth and thus performance. However, adding external caches is not just a simple matter of adding the cache tags and cache RAM because of the issues of cache coherence.

To ensure coherence between the internal and external caches as well as the main memory system, additional information must be provided by the processor on each memory transfer to instruct the external cache on how to handle the data. Decisions such as whether to cache the data, flush it, force an update and so on can be made using this information. These transfer attributes as they are sometimes called are essential to the correct operation of second level caches.

Discrete caches are space consuming, place artificial limits on system speed, prevent multiprocessor designs and are generally underestimated in their design complexity. The biggest restriction with discrete implementations is upgrading the system clock speed. Having said that, many designers are willing to perform this type of work to squeeze the last bit of performance from their systems.

Most of today's M68000 family designs are capable of running different speed processors on the same hardware by simply replacing the processor and retaining the original memory system and peripherals. The insertion of wait states does not degrade the system performance too much. With a RISC architecture, both the processor and its cache memory need upgrading to prevent waitstates on the processor buses from destroying performance. Faster processors need faster caches which need faster memory chips which will become available. The problem comes with the interface logic and buffering that is needed.

Buffering delays of a few nanoseconds start becoming very significant as cycle times become shorter. In addition, the effect of impedance mismatch and capacitive loading, due to design layout and printed circuit board track lengths, start exaggerating the problems. In effect, it will be virtually impossible to speed upgrade such designs without redesigning the board. The only sensible method of solving these problems is to integrate the various functions on to one piece of silicon manufactured in the same technology. With both the processor and its cache memory system using similar fabrication techniques, they can easily track each other as the faster versions become available. A board upgrade simply becomes a processor and integrated cache replacement, without the need for a total redesign.

Conclusions

The most obvious conclusion is that designing a memory system for today's high performance processors is extremely complicated compared with that of a few years ago. This is further compounded by the unfortunate fact that many of the design trade offs are so system and software dependent, it is difficult to draw parallels from other systems. However, there are some general statements which can be made.

CISC processors are the most tolerant of slow memory compared with DSP and RISC processors, although the effect on performance is different in each case. The design technique of inserting wait states but increasing the processor clock generally increases CISC performance but decreases RISC and DSP performance. In practice, RISC processors with two or three wait states for all memory accesses may be a waste of time.

For higher speed performance, cache memories provide an ideal buffer between the processor and slow memory but their complexity must not be underestimated. Special care must be taken to ensure cache coherency and integrated solutions are preferable to discrete implementations.

The inclusion of DRAM, memory management and caches can involve additional 'pseudo-random' delays during memory accesses, caused by memory refresh, page faults and cache flushing, etc. These cause distortion within DSP applications and are usually not to be avoided if possible. They can cause problems within real-time applications running on other systems and these timing differences should be taken into consideration within the design.

7 Real-time software, interrupts and exceptions

What is real-time software ?

Real-time software differentiates itself from many other types of software by its reaction to external stimuli. For software to be described as 'real-time' it must respond in a known maximum time interval to outside stimuli such as interrupts. Many process control applications require this ability to control machinery like conveyor belts, which are fitted with limit switches to prevent objects from falling off. The switch is located such that when it is tripped, the control system can respond and stop the belt before the object being carried falls off the end.

Within such a system, the software must demonstrate realtime characteristics to guarantee that it will respond and switch off the belt within a certain time when the limit switch is tripped. The actual response time can be 3 ms or 300 s, as long as it is guaranteed. In the conveyor belt example, a longer guaranteed response would require positioning the limit switch further up the belt.

The easiest way to obtain realtime characteristics is to use a realtime operating system, like VRTX, OS-9 or pDOS. These operating systems, and others like them, are built around a central real-time kernel which controls the basic processor hardware and provides facilities for software applications to use. The kernels are multitasking, allowing multiple applications to execute simultaneously and respond directly to interrupts etc. With these kernels, there is basically one key time measurement — the context switch.

When a context switch occurs, the currently executing application is replaced by another. This may happen as part of the normal multitasking operation or when an interrupt needs servicing. In the latter case, the servicing application is switched in as part of the system response. A context switch involves saving the processor's registers into an application control block and replacing them with data from another application control block. Using today's processors, context switch times of about 100 microseconds are quite normal.

Responding to an interrupt

There are four general operations involved with generating a realtime response to an external interrupt:

- Interrupting the processor.
- Servicing the interrupt at the kernel level.
- Table searching to locate the service task.
- Performing a context switch to load it.

The speed of these operations is dependent both on the efficiency of the software and on the performance of the processor.

Interrupting the processor

At this stage, the main factor in determining response time is the execution time of each processor instruction. With many of today's processors, an interrupt can only be acknowledged on completion of the current instruction and this execution time can be anything from a two to two hundred clock cycles if complex addressing modes and operations are executed.

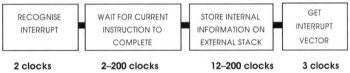

RECOGNISE INTERRUPT	WAIT FOR CURRENT INSTRUCTION TO COMPLETE	STORE INTERNAL INFORMATION ON EXTERNAL STACK	GET INTERRUPT VECTOR
2 clocks	2–200 clocks	12–200 clocks	3 clocks

A typical processor interrupt sequence

Once the processor has been interrupted, it has to store enough information to allow it to return to its activities prior to the interrupt. External memory is often used for this operation, holding the vector table containing pointers to the individual interrupt service routines. At this stage, performance is almost entirely dependent on hardware design, primarily memory access times and processor speeds

Servicing the interrupt

The interrupt service routine is the first software to deal with the interrupt. It operates at a very low level as part of the kernel and is closely associated with the processor itself. The speed of this operation depends both on the efficiency of the kernel software and on the performance of the processor executing it.

The basic flow is simple.The interrupt routine services the kernel's needs and proceeds to locate the task attached to the interrupt. If a timer-generated interrupt is received, the service routine may update the kernel's system clock and perform other housekeeping functions before activating the tasks associated with that interrupt vector. This service software is critical to correct operation and its characteristics determine the response times offered by the operating system. For instance, extensive checking may be carried out to ensure system integrity at the expense of speed. More trusting software may make the opposite trade-off and sacrifice system checking for speed. Once the basic kernel needs have been completed, a higher level routine takes over.

Locating associated tasks

Application tasks running under the kernel's control may be waiting for the interrupt to occur and trigger some task response. This may be the activation of a dormant task to perform some specific routine (e.g. switch the conveyor belt off). The normal method of determining who is interested is extensive table searching. The actual implementation differs between kernels, but the basic operation is similar to that shown.

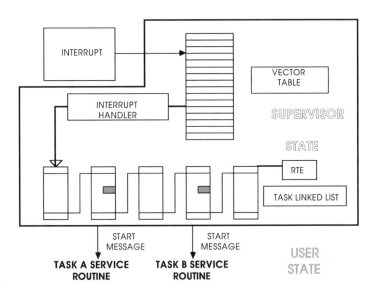

Real-time kernel operation

Various tables or linked lists are interrogated to establish which tasks have declared an interest in the particular interrupt. Each interested party is then informed through a message passed by the kernel's location routine. If multiple tasks are attached, each task is notified in turn, usually on a priority basis. This allows more critical functions to be informed first. Invariably, a context switch is performed to allow a higher priority task to action its message. In our example, if the conveyor belt needs to be switched off, the task that performs it cannot wait until the current task completes its operation. A context switch is required to replace it and allow the new task to switch off the belt. Again, the speed of this operation is dependant on the efficiency of the software and similar trade-off decisions as before.

Context switches

If a task needs to be activated and executed to action the interrupt, the context switch time needs to be added to the previous times to provide the total response time. The time taken to perform a context switch depends on the number of processor registers that require saving, memory access times and the software algorithms used for housekeeping within the kernel. Context switch times quoted by many operating system vendors include the system overheads associated with deciding which task is selected for the next execution time slot. Once the new task is switched, it can take whatever action is required.

For an M68000 processor, interrupt sequences and responses are determined by a combination of the currently executing instruction and the time taken to build the appropriate external stack frame. To generate an interrupt, the level is encoded in binary on to the three interrupt pins. These pins are sampled regularly and the interrupt must be present and steady for a minimum of two system clocks before it is recognised. However, Motorola's specifications recommend that an interrupt level is maintained until the processor acknowledges it. In other words, do not assume that a pulse width of 2 or more clocks will always generate an interrupt — it won't! The processor compares this interrupt level against the interrupt mask stored in the status word. If the incoming level is higher than that of the mask the processor services the interrupt. If the interrupt is of a lower level, it is ignored. A level 7 interrupt is always taken and cannot be masked.

The processor can only service an interrupt on instruction boundaries, i.e. when it has completed executing the current instruction but before the next one commences. As stated previously, this can be a long time if complex addressing modes and instructions are being used. At the boundary, the processor checks the interrupt level status and, if needed, starts to process the interrupt. It switches into SUPERVISOR mode, makes an internal copy of the status register, suppresses any tracing and raises the interrupt mask to that of the external interrupt. This last action is necessary to prevent the same interrupt from repeatedly generating interrupts. The processor then starts an interrupt acknowledgement cycle where it either receives a vector number or assumes it is to use an predefined autovector assigned to that particular level. It then builds the external stack frame comprising of four words. These store the status register, program counter, along with a vector offset and frame identifier. The processor can now locate the service routine from the vector location within the vector table, and several microseconds after the interrupt was recognized, the service routine is started.

After servicing the interrupt, the RTE (ReTurn from Exception) instruction is executed. The information from the stack frame is read back into the processor and normal processing commences.

Improving performance

The M68000 sequence follows the classical interrupt process. The use of an external stack frame allows the nesting of interrupts and other exceptions until the stack runs out of memory. There are two ways in which this sequence can be shortened. The first removes the need to wait for the current execution to complete. However, the corresponding stack frame is larger, due to the extra processor context information that is stored. All the intermediate and internal data, microcode and pipeline contents have to be saved externally. A second method involves the use of a hardware stack where the time consuming stacking operations are performed internally with a far shorter time.

The M88000 and DSP56000 use variations on these ideas to reduce their interrupt latency.

Interrupting CISC and RISC processors

The interrupt process for a RISC processor appears at first sight to be extremely similar to that of a CISC processor except the reponse time is a lot quicker. However, this improvement is obtained by removing much of the work that is carried out by a CISC processor and placing it in the domain of the interrupt service routine. In this way, the programmer can decide how much overhead can be tolerated against other characteristics.

To examine these differences, compare how the MC88100 RISC processor and the MC68000 CISC processor cope with interrupts and other exceptions. In the case of the MC88100, it has a single interrupt pin which is used to signal an external interrupt. If this level sensitive pin is asserted for two successive falling clock edges, an interrupt is internally signalled and the processor context saved to an internal set of shadow registers — not to external memory. In comparison, the MC68000 has three interrupt pins with seven levels of interrupt and the signal is asserted until recognised.

The current MC88100 instruction does not have to complete, unlike that of the MC68000 and the only variable in the latency calculation is the time taken to complete any data and instruction unit memory accesses. Worst cases are four operations for the data unit and two for the instruction unit. A typical value is six clock cycles which, at 25 MHz, is less than 0.25 microseconds. This is at least an order of magnitude faster than more conventional architectures. The diagram shows the sequence for both architectures.

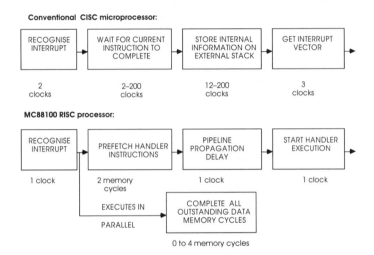

*Interrupting an
MC88100*

Once interrupted, the MC88100 shadow registers are frozen and no further updating can take place. If another interrupt occurs while shadowing is frozen, the processor cannot preserve its context and takes another ERROR exception. Such interrupt routines should therefore execute without exception, or the shadow registers saved to external memory and shadowing enabled.

RISC interrupt service routines

RISC architectures trust software far more than conventional processors and allow far greater control over exception handling. The interrupt service routine is selected by a vector passed by the interrupting device. The vector table, is different from conventional processors in that each vector contains the first two instructions of the handler and not an address pointer. This allows instant execution of the routine.

One of the instructions must jump or branch to the remainder of the routine. Usually, this is a delayed branch which prevents unnecessary pipeline flushing and maintains single cycle instruction execution. If shadowing is re-enabled, further interrupts and exceptions can be nested. The decision to allow this is totally under software control. Software can decide if it wishes to trade system checking for speed.

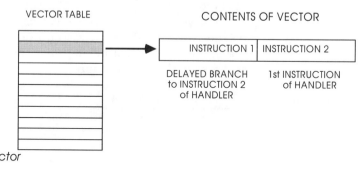

VECTOR TABLE   CONTENTS OF VECTOR

| INSTRUCTION 1 | INSTRUCTION 2 |

DELAYED BRANCH   1st INSTRUCTION
to INSTRUCTION 2   of HANDLER
of HANDLER

The MC88100 vector table

Vector Offset (hex)	Exception	
0 0000	Reserved	
0 0100	System Reset	Power-on, Hard & Soft Resets
0 0200	Machine Check	Eabled through **MSR**[ME]
0 0300	Data Access	Data Page Fault/Memory Protection
0 0400	Instruction Access	Instr. Page Fault/Memory Protection
0 0500	External Interrupt	$\overline{INT}$
0 0600	Alignment	Access crosses Segment or Page
0 0700	Program	Instr. Traps, Errors, Illegal, Privileged
0 0800	Floating-Point Unavailable	MSR[FP]=0 & F.P. Instruction encountered
0 0900	Decrementer	Decrementer Register passes through 0
0 0A00	Reserved	
0 0B00	Reserved	
0 0C00	System Call	'sc' Instruction
0 0D00	Trace	Single step instruction trace
0 0E00	Floating -Point Assist	A floating point exception

The basic PowerPC vector table

If speed is chosen, software must be written so that no further exceptions occur. The code must be readily available and not swapped out to disk, stack and memory resources must be adequate, further interrupts require masking and memory systems, especially those with error detection and correction, must not return an error.

Once an exception has been recognised, the program flow changes to the associated exception handler contained in the vector table.

The next stage involves executing the kernel's routines to decide what action is required.

Improving software performance

RISC processors achieve their greater performance by simplifying instructions so that they execute in a single cycle and creating the more conventional complex instructions as series of RISC instructions. Although multiple instructions are executed, overall execution time should be shorter. However, the more complex and powerful the RISC instructions that are executed while still maintaining single cycle execution, the more work the processor performs. The M88000 instruction set has been optimized rather than reduced. This means that many conventional CISC instructions can be replaced on a one-to-one basis, which helps to keep code expansion ratios as close to this ratio as possible. This rich instruction set has tremendous importance, when considering how fast the kernel software routines executes. Three important areas are table indexing and addressing, data fetching and condition testing.

Addressing data

Unlike many RISC architectures, which have only one address mode and require addresses to be calculated prior to every external access, the M88000 instruction set supports five. Probably the most useful is the register indirect with an index, which can be scaled. This allows a pointer to be assigned to a head of a table or set of tables and allows other registers to index into it, dramatically reducing the number of address calculations and instructions required. Simply changing the main pointer register to that of another table allows rapid searching of tables and linked lists. This reduces the number of instructions needed per function and increases throughput.

Fetching data

Having located the data, it now requires fetching. The M88000 is based on a load store architecture, which does not operate directly on external memory — data is loaded into a register, modified and written out again. This effectively removes memory access time from instruction execution time. Such operations take only three instructions and, therefore, three cycles to complete. Direct memory modifying architectures take three instruction times, plus the time for two memory accesses.

An inefficient RISC architecture:

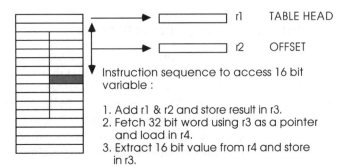

Instruction sequence to access 16 bit variable :

1. Add r1 & r2 and store result in r3.
2. Fetch 32 bit word using r3 as a pointer and load in r4.
3. Extract 16 bit value from r4 and store in r3.

- uses 3 instructions and four registers.

The MC88100:

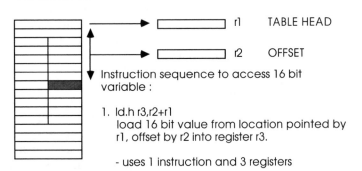

Instruction sequence to access 16 bit variable :

1. ld.h r3,r2+r1
   load 16 bit value from location pointed by r1, offset by r2 into register r3.

- uses 1 instruction and 3 registers

Accessing data within memory tables

Further consideration must be given to how much data can be fetched and how this is performed. Most compiler-generated data, like 'C' variables such as char, short and int, are 8 or 16 bits in length, yet current RISC architectures are 32 bit in data size. While many RISC architectures can deal with the corresponding byte and halfword data sizes internally, they have to fetch such data using a multiple instruction sequence. The first instruction calculates the address and loads it into the address register, a 32 bit wide word of data is fetched into a register and, finally, a third instruction is needed to extract the relevant byte or halfword from that register and store it in yet another register. The MC88100 has four data byte enable signals which allow it to fetch individual bytes and halfwords directly in a single cycle. This again reduces the code expansion, the number of cycles needed to perform the task, and, ultimately, improves system perform-ance.

The combination of these two factors can reduce the number of instructions needed to access a byte of data from a table by a factor of three, compared to other RISC architectures. In the example shown, the MC88100 can access data directly by fetching a byte from a table pointed to by a register and indexed by a second. With a poor RISC instruction set, the address is calculated and loaded into a register. A 32 bit value is then loaded and, finally, the byte is extracted, making threeinstructions in total. This operation is repeated many times by a kernel — savings in this area are essential to maximising performance. With the data located and loaded, the kernel must now test it.

Testing data

Conventional processors have a condition code regis-ter which is usually updated at the end of every instruction, allowing easy condition testing. Various combinations of bits within the register indicate if the certain conditions exist.

Maintaining this register imposes a large overhead on the processor and it is of no surprise that the M88000 family, like most RISC architectures, does not have a specific condi-tion code register. Instead, any general purpose register can be temporarily used as a condition code register.

To compare two values held in two registers, the MC88100 compare instruction can be used. This tests for all conditions and sets appropriate bits within a third specified register. The next instruction is a branch on bit set or clear, which redirects program flow as necessary. With further bit masking etc., extremely complex conditions can be tested for. To test for five particular conditions only requires six instruc-tions — a compare followed by five branch on bit condition instructions. In addition, delayed branch versions of the in-struction are available, which improve efficiency by remov-ing pipeline stalls caused by flushing.

However, the MC88100 has another, more efficient, method of branching on condition. The previous example compared two values and needed a two-instruction sequence. There is a branch on condition instruction which compares, tests and branches in a single instruction and, with the de-layed slot version, executes effectively in a single cycle. This instruction compares a single value with zero and has an interesting effect on compiler and operating system software.

Instead of counting upwards within loops, it is more efficient to count down to zero. A high level language procedure may require a loop variable starting at 0, incrementing by 1 and causing termination at value 10. This causes ten iterations of the loop. The compiler, for maximum efficiency, should code this as a loop variable initialised with 10, decrementing by 1 and terminating at 0. This type of optimisation can only be performed if the loop variable is of limited scope and not referenced elsewhere. For maximum performance, loops and other counters should count down rather than up.

Saving and restoring register sets

Having tested the data, the decision to perform a context switch has been taken. The context switch simply saves the current processor register contents and replaces them from memory. The time taken is dependent on the number of registers and the number of memory accesses to save / restore each register. RISC architectures require more registers than other architectures, especially to hold data locally, which other architectures would manipulate directly in memory. A cause of concern with many of today's processors is the large number of registers and the number of clock cycles requires per memory access. However, with reference to context switching times of real-time operating systems, actual register swapping is a small part of the overall time. To save 20 registers, at three clocks per register, takes only 60 clocks. At a 20 MHz clock rate, this is equivalent to 3 microseconds of processing time. With a typical context switch of 100 microseconds and four save or restore operations during the kernel processing, this is still only equivalent to about 12% of the total time.

With the register intensive characteristics of RISC, caused by no direct data manipulation in memory, larger register sets are necessary to maintain performance. However, they require storing during context switching and this potentially increases context switching times. The MC88100 has only 32 general purpose registers, which it can save in 32 clocks. This is still 50% less than conventional processors. Decreasing the number of registers can decrease this time, but any such decision must be carefully considered. If software restricts register usage too much, increased stacking operations may result, which actually make the kernel, application execution and processing less efficient and therefore slower.

Interrupting the DSP56000

Interrupting the DSP56000 from external sources is different from both the M68000 and M88000 processor families. External interrupts normally generate a fast interrupt routine exception. The external interrupt is synchronised with the processor clock for two successive clocks, at which point the processor fetches the two instructions from the appropriate vector location and executes them. Once completed, the program counter simply carries on as if nothing has happened. The advantage is that there is no stack frame building or any other such delays. The disadvantage concerns the size of the routine that can be executed and the resources allocated. When using such a technique, it is usual to allocate a couple of address registers for the fast interrupt routine to use. This allows coefficient tables to be built, almost in parallel with normal execution.

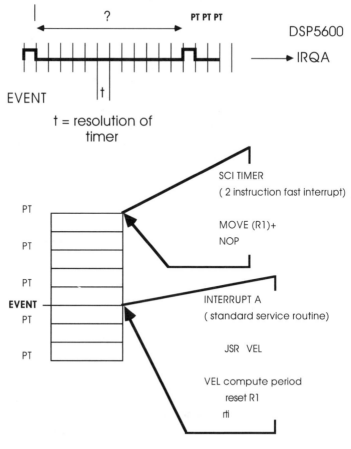

Using interrupts to time an event

The SCI timer is programmed to generate a two instruction fast interrupt which simply autoincrements register R1. This acts as a simple counter which times the period between events. The event itself generates an IRQA interrupt, which forces a standard service routine. The exception handler jumps to routine VEL which processes the data (i.e. takes the value of R and uses it to compute the period), resets R1 and returns from the interrupt.

Diadic versus triadic instruction sets

One major difference between the M68000, the DSP56000 and the M88000 and PowerPC family instruction sets is the number of source and destination addresses that each instruction can specify. The M68000 family uses a diadic set with two components forming two data sources and the instruction results being returned in one of the sources. The M88000 RISC processor uses a triadic model with two sources and a third, separate destination. The DSP56000 is essentially a diadic model but has some instructions that can perform parallel moves with extra addresses. What are the advantages?

The obvious advantage concerns the preservation of data within the register set or in memory. The diadic instruction destroys the contents of one of the sources on its completion, which either forces the data to be fetched once more, or inserts an instruction to copy the destroyed sources's contents to the stack or another register, prior to the instruction. Compilers often generate an instruction pair which can be replaced by a single triadic instruction. In this case, the M88000 reduced instruction set, with its triadic operations, is actually more efficient than the complex instruction set of the M68000. However, encoding three operands into a single fixed length instruction can restrict the number and different types of operations that can be executed. With a 32 bit op code and 32 registers, the M88000 uses 15 bits for register encoding and leaves 17 bits to describe the operation or for immediate values etc. With a register set of 256 registers, there are only 8 bits left to encode the operation etc. With their single cycle execution, RISC processors must have a fixed op code length and cannot do as the M68000 and DSP56000 families, i.e. simply extend the op code as necessary.

CISC processors, with their memory to memory architectures which need variable length instructions to accommodate direct addresses with scaling etc., have an average instruction length of 24 to 32 bits rather than the minimum 16 bit op code. Comparing lengths between MC68020 and M88000 code gives a typical expansion of about 30%. This gives a good indication of how powerful the M88000 instruction set actually is. With its greater power, it performs more work per clock cycle — i.e. its MIPS are worth more than other architectures!

Instruction restart versus instruction continuation

Another aspect of real-time performance concerns the mechanism used to restore processor context during exception processing caused through a bus error or a page fault. Instruction restart effectively backs up the machine to the point in the instruction flow where the error occurred. The processor re-executes the instruction and carries on. The instruction continuation stores all the internal data and allows the errant bus cycle to be restarted, even if it is in the middle of an instruction.

The continuation mechanism is undoubtedly easier for software to handle, yet pays the penalty of having extremely large stack frames. The restart mechanism is easier from a hardware perspective, yet can pose increased software overheads. The handler has to determine how far back to restart the machine and must ensure that resources are in the correct state before commencing.

While the majority of the M68000 family are of the continuation type, the MC68040, the M88000 and PowerPC families are of the restart type. As processors increase in speed and complexity, the penalty of large stack frames shifts the balance in favour of the restart model.

External memory and real-time performance

While all RISC systems should be designed with single-cycle memory access for optimum performance, the practicalities are that memory cycles often incur wait states or bus delays. Unfortunately, RISC architectures cannot tolerate such delays — one wait state halves the performance, two reduces performance to a third. This can have a dramatic effect on real-time performance. All the advantages gained with the new architecture may be lost.

The MC88200 CMMU can allow slower and more cost effective memory to be used, without the performance penalties. It provides wait state free cache memory, which prevents such delays from affecting the MC88100 CPU performance. The MC88100 can save its registers into the MC88200 cache memory on a single cycle basis, irrespective of how slow the external memory subsystem is. In addition, critical software routines can be cached in another CMMU on the instruction bus, again allowing single cycle access and execution. If instructions have to be fetched from dynamic memory, with typical times ranging from four to forty clocks, the kernel execution time becomes the dominant factor within the context switch time. Again, any advantage offered by RISC may be lost through a poor memory subsystem. Similarly, the PowerPC processors with their integrated caches can utilise the same techniques to improve their performance.

However, there are some potential penalties for any system that uses caches and memory management which must be considered.

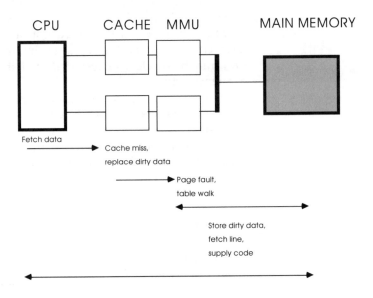

The impact of a cache miss

Consider the system in the diagram. The processor is using a Harvard architecture with combined caches and memory management units to buffer it from the slower main memory. The caches are operating in copyback mode to provide further speed improvements. The processor receives an interrupt and immediately starts exception processing. Although the internal context is preserved in shadow registers, the vector table, exception routines and data exist in external memory.

In this example, the first data fetch causes a cache miss. All the cache lines are full and contain some dirty data, therefore the cache must update main memory with a cache line before fetching the instruction. This involves an address translation, which causes a page fault. The MMU now has to perform an external table walk before the data can be stored. This has to complete before the cache line can be written out which, in turn, must complete before the first instruction of the exception routine can be executed. The effect is staggering

— the quick six cycle interrupt latency is totally overshadowed by the 12 or so memory accesses that must be completed simply to get the first instruction. This may be a worst-case scenario, but it has to be considered in any real-time design.

This problem can be contained by declaring exception routines and data as non-cachable, or through the use of a BATC or transparent window to remove the page fault and table walk. These techniques couple the CPU directly to external memory which, if slow, can be extremely detrimental to performance. Small areas of very fast memory can be reserved for exception handling to solve this problem; locking routines in cache can also be solutions, at the expense of performance in other system functions.

Register windowing

Another architectural feature often put forward as a good way of improving processor context switch and procedural call times is the idea of register windows. A small area of onchip memory is divided into register banks which can be switched to store the old contents and give a new clear set without having to store the data on a an external stack frame.

This solution has one drawback — what happens when all the windows have been used? The answer is that data must be saved out to memory to free up a window. The next question concerns which window is saved. In practice, the whole set of windows is transferred out with the inevitable transfer delays. A typical number may be about eight windows, which may or may not be enough to support an operating system call. For real-time applications, the worst case scenario of a complete window set transfer must be assumed.

Combining architectures

The most obvious method of improving real-time performance within embedded control systems is to increase the processing power. Faced with the task of sending and receiving serial data at 1 Mbit/s, or detecting missing teeth on a gear wheel to generate ignition timings at 7,000 rpm, this appears to be the most sensible choice. Such systems generate a lot of interrupts and require fairly intensive data processing to perform serial data checking or timing measurements between these events. Where microprocessors have been used to implement such designs, the system has centred around a single processor controlling standard timer and serial comms peripherals. This approach started with 8 bit microprocessors and, as demands have increased, faster generations have simply been slotted in to solve the problem.

With the advent of microprocessors like the Motorola M88000 and the DSP56000/DSP96000 digital signal processors augmenting the M68000 family, a designer has several options to solve the problem. The most obvious is to select the fastest, most performant processor available and use its increased power to reduce interrupt latency and increase data processing. However, each of these families has its strengths and careful consideration must be made of system needs.

As the processor increases in power, demands on the memory system increase and, as previously described, cache systems can be used — they, too can introduce further difficulties.

An alternative approach to this problem is to adopt a multiprocessing system where extra processing is distributed throughout the system. This has been used in many VMEbus systems, for example. Rather than use a single higher cost processor, several lower cost processors are used and the workload is distributed across the system.

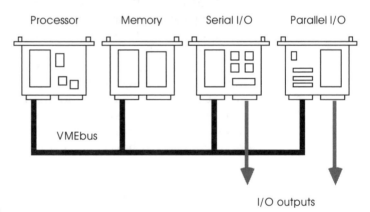

A VMEbus based system

In this configuration, serial and parallel I/O is controlled by separate processors located on each I/O board. With this intelligence, the overhead of servicing interrupts etc. is handled onboard and frees up the main processor to perform other tasks. The onboard processors may not be the same as the main CPU, but instead can be matched to have the right characteristics required by the interface software to control the I/O.

This solution is often not possible for many implementations, due to the higher component count costs involved with designing the bus interfaces, the processor to processor communication, bus arbitration, etc. However, silicon technology which has increased clock speeds and device density to provide today´s high performance processors can equally be used to integrate a distributed processor system on to one chip, with the resultant reduced chip count and costs.

The M68300 family

There are now several members of the family currently available. They combine a M68000 family processor and its asynchronous memory interface, with all the standard interface logic of chip selects, wait state generators, bus and watchdog timers, into a system interface module and use a second RISC-type processor to handle specific I/O functions.

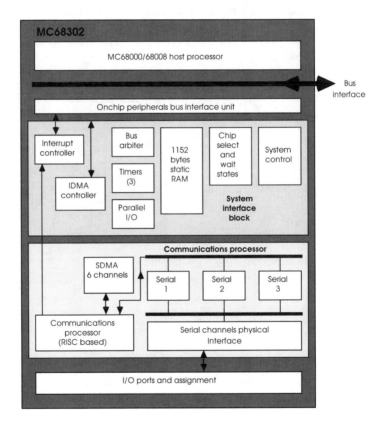

*The MC68302 integrated
multiprotocol processor*

The MC68302 uses a 16 MHz MC68000 processor core with power down modes and either a 8 or 16 bit external bus. The system interface block contains 1,152 bytes of dual port RAM, 28 pins of parallel I/O, an interrupt controller and a DMA device, as well as the standard glue logic. The communications processor is a RISC machine that controls three multiprotocol channels, each with their own pair of DMA channels. The channels support BISYNC, DDCMP, V.110, HDLC synchronous modes and standard UART functions. This processor takes buffer structures from either the internal or external RAM and takes care of the day-to-day activities of the serial channels. It programs the DMA channel to transfer the data, performs the character and address comparisons and cyclic redundancy check (CRC) generation and checking. The processor has sufficient power to cope with a combined data rate of 2 Mbits per second across the three channels. Assuming an 8 bit character and a single interrupt to poll all three channels, the processor is handling the equivalent of an interrupt every 12 microseconds. In addition, it is performing all the data framing etc. While this is going on, the onchip M68000 is free to perform other functions, like the higher layers of X.25 or other OSI protocols as shown.

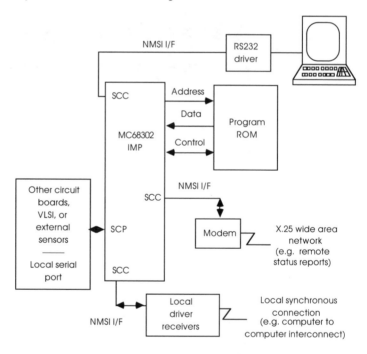

A typical X25-ISDN terminal interface

The MC68332 is similar to the MC68302, except that it has a CPU32 processor (MC68020 based) running at 16 MHz and a timer processor unit instead of a communications processor. This has 16 channels which are controlled by a RISC-like processor to perform virtually any timing function. The timing resolution is down to 250 nanoseconds with an external clock source or 500 nanoseconds with an internal one.

The timer processor can perform the common timer algorithms on any of the 16 channels without placing any overhead on the CPU32.

A queued serial channel and 2 kbits of power down static RAM are also onchip and for many applications, all that is required to complete a working system is an external program EPROM and a clock.

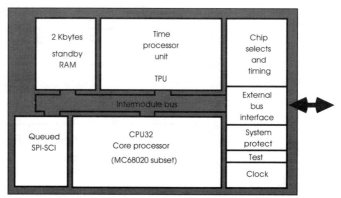

*The MC68332 block
diagram*

Software considerations

Another benefit of these highly integrated distributed processing chips is the reduction in software that is needed to be written. For a more traditional communications controller, the low level interrupt handlers, framers, etc., can be extremely difficult to write, especially with the tight timing restraints often encountered. To program the MC68302 comms controller, a memory-based buffer structure is used which consists of a frame status, a data counter and a data pointer to the receive and transmit data resident in memory. This can simply be defined as a structure in C. There is no low level bit manipulation, interrupt handling, data checking, etc. This is already handled by the other processor and its embedded code.

Similarly with the MC68332. To perform the missing tooth calculation for generating ignition timing, a set of parameters is calculated by the CPU32 and loaded into a parameter block which commands the timer processor to perform the algorithm. Again, no interrupt routines or periodic peripheral bit manipulation is needed by the CPU32.

The M68000 instruction set compatibility shown by both processors has other advantages — compilers, applications and real-time operating system are freely available, allowing the easy migration or development of code.

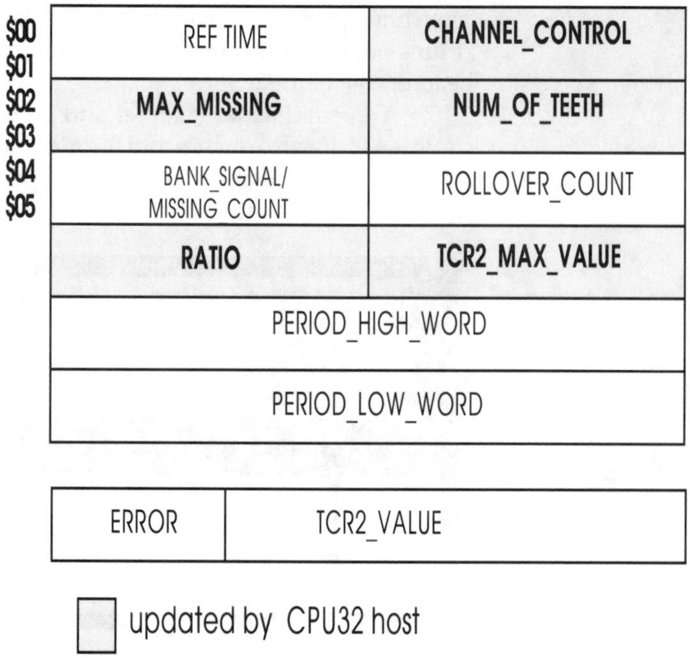

$00 $01	REF TIME	**CHANNEL_CONTROL**
$02 $03	**MAX_MISSING**	**NUM_OF_TEETH**
$04 $05	BANK_SIGNAL/ MISSING_COUNT	ROLLOVER_COUNT
	RATIO	**TCR2_MAX_VALUE**
	PERIOD_HIGH_WORD	
	PERIOD_LOW_WORD	

ERROR	TCR2_VALUE

☐ updated by CPU32 host

*The parameter block
for a period measure-
ment with missing
transition detection*

Combining DSP processors

This technique need not be confined to integrated processors. Both the DSP56000 and DSP96000 processors can be configured to act as subordinate processors to an MC68000 or similar type of processor. This is easily performed using their host interface ports, which act as memory windows into the processor, allowing processor-to-processor communication via memory and interrupts. The interface logic is minimal and thus meets the low component requirements of many applications.

Why use a DSP chip? It combines good processing power with internal memory and has extremely quick interrupt latency, typically a couple of clocks for a fast interrupt. This is achieved through the use of a hardware stack and the absence of automatic stack building. The device is highly integrated, with parallel and serial I/O channels, onboard memory, memory interface logic, etc. The register model has general purpose address registers and a dual accumulator data model and, although a little different to an M68000 or M88000, is capable of high level programming in C. The processor throughput is good — in 75 nanoseconds, the DSP56000 can perform a multiply accumulate and two data transfers simultaneously. The DSP96000 can also do this but performs the operations in IEEE floating point precision and calculates differences as well. In MIPS figures, these processors are performing at about 13 MIPS and peaking at over 40 MIPS.

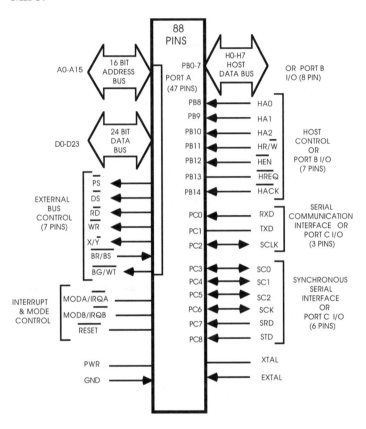

The DSP56000/56001 processor

In the typical configuration shown, the DSP56001 loads its program from external EPROM and then executes from internal memory, not needing external memory. The program/data can be altered or downloaded by the M68000 passing data via the host interface, forcing the DSP56001 to execute a fast interrupt. There are 16 routines that can be selected by the M68000 and they effectively force the DSP56001 to execute a simple two-instruction sequence with no interrupt overhead. Using this technique, 24 bits of data can be loaded into the host interface receive registers and a fast interrupt can be used to force the DSP to load the data into its memory. A sequence of such transfers can be used prior to redirecting the DSP56001 to another set of data or routine.

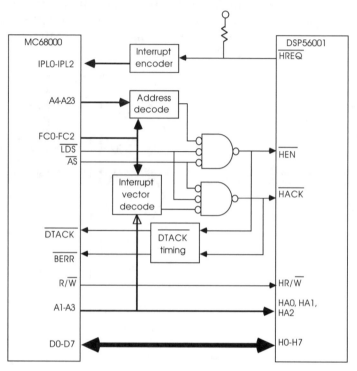

*An MC68000/DSP56000
multiprocessor
configuration*

This simple combination provides very fast interrupt handling and processing from the DSP56001 and good general purpose processing from the M68000. Multiple DSP processors can be controlled in this way to create arrays. Typical examples are adaptive filtering for digital audio, suspension systems and graphics accelerators. The NEXT computer, designed by Steve Jobs, has used this principle, coupling an MC68030 and a DSP56000 together. The MC68030 runs UNIX while the DSP56000, by virtue of different algorithms and control, provides the system with a modem, fax machine, CD quality sound and speech synthesis and recognition. To implement these facilities with discrete peripherals would place a tremendous overhead on the processor. While a faster processor, like the MC68040, could cope with that burden, the DSP56000 dispenses with the peripherals and removes the burden at a much lower cost.

An application example describing hard disk motor control in Chapter 9 uses this technique.

Conclusions

RISC, CISC and DSP processors all have different methods of handling interrupts and exceptions which, in turn, have their various trade-offs. When considering the whole aspect of interrupt latency and context switching, it is important to examine the complete process involved, from the interrupt generation through to performing the required service. It is all too easy to blinker the process to functions which are thought to be critical when other areas are impacting performance far more. While the lure of the latest and most powerful processors is extremely attractive, multiprocessing or integrated systems may actually achieve the needed goals with less effort and cost.

8 Multiprocessing

The use of multiple processors to improve performance within a system has been common for many years but, with the advent of fast processors, memory subsystems and buses, there has been an almost unparalleled interest in more recent years. Again, there are many differing types of arrangements and considerations which need careful consideration. The most important aspect of any multiprocessor design centres around the nature of the task(s) to be performed and the method of communication used.

While it may be easy to believe the hype associated with some of the larger multiprocessor systems; i.e. that performance increases linearly with the number of processors, the reality is totally different. If the communication overhead is greater than the performance gain, the system will plateau and adding extra processing units will result in minimal performance gain. Gene Amdahl, who designed IBM's mainframes before his own computers, expresses this relationship as :

$$N/\log N$$

where N is the number of processors. Many multiprocessor systems make several assumptions about the type of software program that is going to be executed and this often limits their use to niche applications. In other words, a 100 CPU parallel processor computer is unlikely to run a spreadsheet or wordprocessing application any faster than a single processor system but will be able to model weather systems. The difference depends on the nature of the application — a spreadsheet has serial data and program execution and cannot subsequently be broken into components which can be executed in parallel. An environmental model can.

Every processor has a separate data and instruction stream associated with it and this gives four main types of processor design, depending on whether multiple or single streams are used.

SISD — single instruction, single data

The single instruction, single data stream machine is the most commonly implemented design and was the basis of many early mainframe and super computer designs. All the processors described so far, ranging from the M6800 through to the M88000 series, are essentially single instruction, single data machines. This definition can include pipelined arrangements, multiple functional units and Harvard organization. The basic architecture consists of two independent streams supply the instructions and data to the processor unit within the shaded box. The three-stage pipeline processes the data according to the instruction and stores the results or performs an I/O function. This type of machine is essentially a serial execution machine, although multiple resources and overlapping may be used to reduce the time taken to perform each instruction.

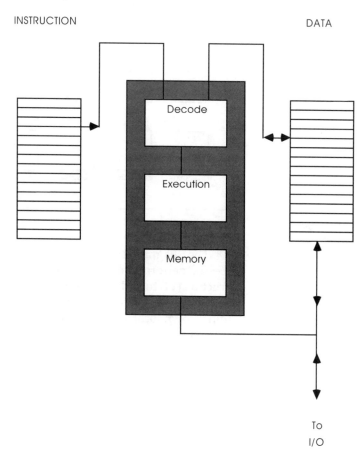

INSTRUCTION

DATA

Decode

Execution

Memory

To
I/O

Basic SISD architecture

SIMD — single instruction, multiple data

In this configuration, used in array processors, one instruction stream is used to process multiple data streams using many processing elements. A single instruction is received at a time but may be conditionally executed by the individual processing units on their own individual data, stored in local memory. This conditional type of execution is necessary for systems or program sequences which require serial execution and cannot make use of parallelism. The communication methods used are based around ring, mesh and broadcast networks.

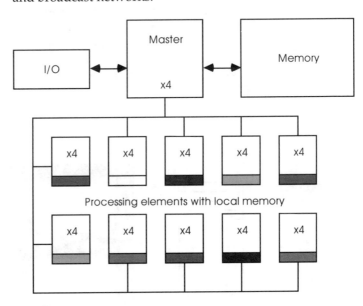

Basic SIMD architecture

The diagram shows the basic principles. A master processor performs the basic I/O functions and issues the instruction (x4) to each processing element via the communication network. Each element then processes the data stored within its own local memory (patterned box). This process is repeated until all the data has been successfully processed. At this point the results are passed to the master for consolidation and further data dispersed through the network, ready for the next set of processing.

This data collation and dispersion adds system overhead and increases dramatically as the number of elements goes up.

MIMD — multiple instruction, multiple data

The multiple instruction, multiple data machine is the traditional multiprocessor system which can execute multiple programs operating on multiple data streams. Each instruction stream has its own data stream which may come from shared or local memory. Processor-to-processor communication can be based around shared memory, serial networks and other links, like crossbar switches and multiport memory. The diagram shows a typical example using a VMEbus system.

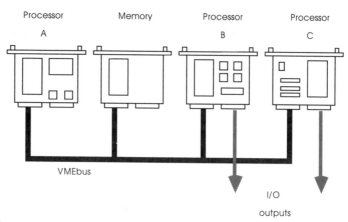

An example MIMD system

MISD — multiple instruction, single data

This design is essentially hypothetical as there are no real implementations, although the execution pipeline fits the description. Each stage of the pipeline acts as an independent processing unit that operates on the data passing through it and before passing it on to the next stage in the chain.

Constructing a MIMD architecture

This type of architecture is the most often implemented with microprocessors and, with the advent of fast processors with multiprocessing support and reduced external memory bandwidth, is becoming increasingly attractive for designers to implement.

The basic communication method involves the use of shared memory, which is used for passing data, program and control information between the processors. Access to such areas is controlled using a semaphore technique implemented with an locking instruction such as the M68000 TAS (test and set) instruction, the M88000 xmem instruction, or with the reservation method used in the PowerPC architecture. This allows each processor to handshake for access, preventing simultaneous access and the system corruption it produces. Such configurations are easy to design but careful consideration must be made of the bandwidth needs and memory organization.

It is essential that each processor has its own local memory and that the bulk of its processing uses this memory, thus reserving the shared memory solely for message passing. The amount of message passing determines the number of processors the system can support. For practical reasons, such systems usually have a central I/O system for mass storage and serial comms etc. This creates a single resource which competes with the message passing traffic and reduces the number of supported processors.

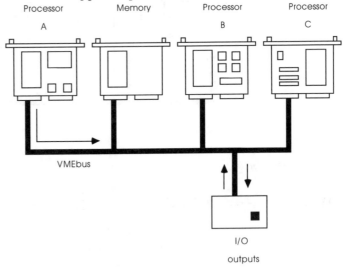

The solution is to take the I/O traffic off the message passing bus. A variation on this design is to use part of the dual ported memory on the processor boards as the message area. This again appears to be desirable on first inspection but can greatly reduce performance. Most 'dual port' memory systems are constructed using a shared memory technique where two or more processors can access the same memory — one at a time. The difficulty arises when the message passing area is used by another processor.

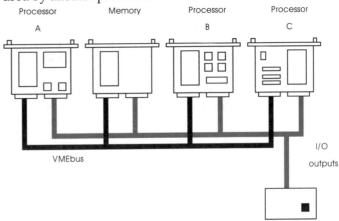

It obtains access and effectively prevents the the local processor accessing its own local memory. Although message passing is not stopped, the local processing is halted and thus the system loses performance. In such configurations, local memory effectively appears on the same bus as the message passing area and local processing consumes message passing bandwidth. For a system where message passing bandwidth is crucial to the design success, this is clearly unacceptable.

The key to such designs lies in optimising the memory interface so that the memory hierarchy is matched to the needs of the various levels within the system.

Processor bandwidths

On of the key restrictions with processors in a multiprocessor design is the external bandwidth that they need. The M68000 family has steadily decreased its external bus bandwidth requirements by the adoption of internal caches, reducing the number of clocks per memory access and through the use of burst filling.

	MC68000	MC68020	MC68030	MC68040
Clocks per Cycle	4	3	2	2
Bus Bandwidth	80%	60%	50%	30%
Nominal Clock Frequency	8	16	20	25

The M68000 memory requirements

The bus bandwidth figures refer to the processor's needs and indicate that the more advanced parts are more suitable than their predecessors for multiprocessing systems. These figures do not take into account the beneficial effect of local memory and caches.

The histogram shows the benefit of local memory. The vertical axis shows the amount of global or shared memory access for a given amount of total memory access. The data for the bus bandwidth figure of 80% is the previous table. The graph shows that large amounts of local memory dramatically reduce the bus bandwidth requirement. To achieve an overall external bus bandwidth of about 20%, enough local memory must be present to meet about 75% of the processor's need.

A comparative bandwidth figure for a RISC processor with reasonable caches — an MC88100 with 2 MC88200 or a MPC601 or MPC603 — is about 25%. The figures for a DSP device like the DSP56000 are harder to calculate. If the program and data can be stored internally, the external bus needs are effectively zero. If external memory is used, the figure becomes 100%. A typical figure is in between and totally software dependent.

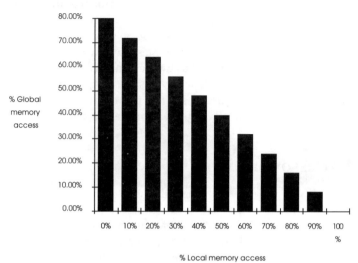

% Local memory access

M68000 memory access characteristics

These figures indicate the number of processors which can effectively be linked in a multiprocessor system, while sharing a single resource or memory. For both the MC68040 and MC88100/MC88200 configurations, this boundary is reached at about four processors. To go beyond requires additional memory resources, like secondary caches and local memory, or the use of other communication techniques, such as serial networks, cross bar switches, etc. The question immediately arises of how to profile such a system and therefore know how to make the trade-offs.

By profiling a system, a good understanding of how the individual processors within a MIMD machine interact and use resources is derived. The basic techniques involve breaking down the overall system into its individual elements and establishing each individual element's needs.

Profiling

First, determine the basic level of performance needed to carry out the given task — i.e. processing throughput, memory needs and speeds, any time critical operations and the amount of data access versus computation.

Frequently, this type of information can only be obtained from running specific benchmarks and code sequences, and from a good system knowledge. In the past, real hardware was often used to provide this data. While providing definitive data on how fast, given a certain configuration, it was impossible to determine the number of data accesses, branches taken, cache hit ratios, etc. With the advent of simulators, this type of data is easily available and is worth its weight in gold. The Oasys M88000 simulator can run code and provide the number of instructions executed, the different types, MMU ATC cache hits and misses, external cache hit and miss ratios and the number of data and instruction fetches performed. This type of analysis through simulation is becoming a standard part of the support tools available for a processor. The DSP56000 simulator can perform a similar function and is also capable of simulating a multiprocessor system. Armed with this type of information, the system profile can be determined and analysed.

Cost of memory access

All processor systems use differing types of memory within their design, depending on the costs, access times and density needed. The hierarchy shows some interesting characteristics.

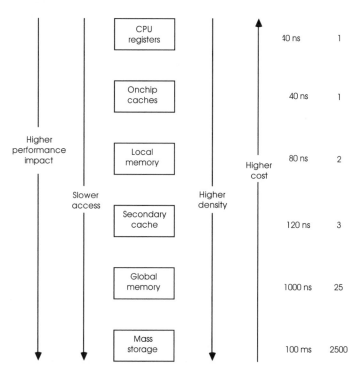

The memory hierarchy

The fastest access is undoubtedly using processor registers or onchip caches. Moving down the hierarchy, increases access time, increases density and decreases the cost per bit. However, if the main processor access is to lower levels, there is a far higher performance impact. Cache memories help by restricting the number of memory accesses the processor has to perform to the levels below the cache. The cache does not help on the first access, but by copying data, provides a benefit on any subsequent accesses. The figures on the right hand side give an indication of the speed differential. CPU register access is some 2,500 times faster than mass storage! The system profile can be matched to provide the right trade-offs.

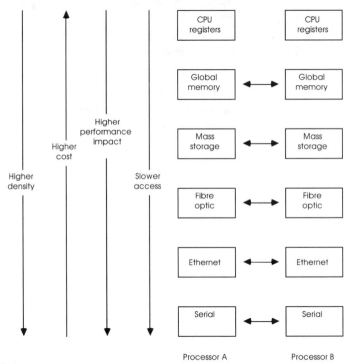

A communication link hierarchy

If a system needs very fast data access, allocating storage on a hard disk greatly impacts performance. In reality, the storage is allocated as close to the processor as possible. If the profile shows that the majority of system functions are computation bound, the ideal is to move code storage up to local memory or onchip cache and move variable storage down. If system spends much of its time page swapping, an increase in local or global memory results in tremendous gains in performance. It can be seen that a good understanding of systems and what they do allows the optimum design to be built.

A similar approach can be done for a multiprocessing system where the points of communication replace the storage needs.

The hierarchy for two processors has other possibilities. Communication links are possible at the register level, however, this type of transfer can only be implemented at a VLSI level and has been left out. Standard communication at the global or shared memory is the fastest available but mass storage using shared files or pages and various speed serial networks can also be used, albeit with greater overhead.

Comparing this diagram with its predecessor, the fastest communication network (i.e. global memory) is about halfway up the memory hierarchy. To reduce the communication overhead further, the ideal solution would be to couple caches together, allowing the extra speed of cache memory to be used.

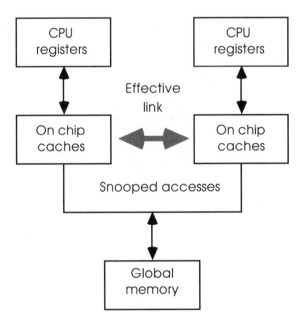

A bus snooping link

This can be done using a bus snooping mechanism. These mechanisms were described in the previous chapter, with reference to cache and data coherency. Using bus snooping to aid processor to processor communication is the same principle viewed from a different aspect. In the example, the processors themselves are responsible for ensuring that shared data is updated correctly — this effectively couples the processor with slow memory and impacts performance. By using the cache, the processor is buffered and only accesses slow memory when the data has been changed. This creates an effective link between the two caches, high up in the hierarchy, for maximum performance.

Shared memory schemes can be implemented with almost any high performance CISC or RISC processor such as any M68000, M88000, PowerPC or DSP56000 processor. All that is needed is some multiport memory and semaphore signalling, using either bit testing or register to memory swaps, to provide control. Links using caches require onchip caches and, for higher performance, bus snooping protocols. The MC68020, MC68030 and MC68040 all have them but only the MC68040 has snooping protocols. The M88000 and PowerPC architectures provides both the cache function and bus snooping necessary for their processors. The DSP56000 does not have any cache but can use its host interface port to provide a fast memory to memory link. It does, however, have a fast synchronous serial interface which can allow data transfers of several megabits per second. Using it, arrays can easily be built.

Fault tolerant systems

Fault tolerant systems are a special type of multiprocessor design, where the goal is increased availability rather than higher performance. Such systems can tolerate component and system failure, albeit with degradation of overall system performance. The basic approach is to use redundancy so if one component fails, there is a backup immediately available to take over. While it is relatively easy to provide uninterruptable power supplies, error detecting and correcting memory, mirror disks etc., to prevent a peripheral failure from bringing down the system, there are several challenges in the processor design.

The biggest problem concerns how the faults are detected and the amount of work necessary to restore the status quo. As with memory and communication links, there is an effective hierarchy for fault detection and backup.

Any fault tolerant system must have multiple processors. A heirachy for a two-processor system and the points within the system where comparison and checks can be performed to detect any failures is shown in the diagram. As the fault detection point moves down, the cost of detection decreases, but system granularity (i.e. the amount of work that may be suspect) increases, as does the amount of restoration work needed.

One detection method is the comparison of data files stored on a hard disk. This is easy to implement but the main processor must effectively be stopped while the comparison is made and the amount of work needed to correct faults is large. The backup process is a byte for byte comparison and is relatively slow due to increased access times. Backup copies of all data files are maintained to provide a known reference in case of a fault, so the system can take the known good backup and simply repeat the processing on another processor. Again, performance impact is large.

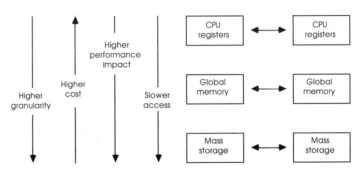

Fault tolerant hierarchy

Comparing memory structures is another technique. Here, memory is split into four separate entities which are accessed by two processors. As with the file comparison method, two of the memory areas are used to hold two copies of a known reference point, while the other two are usable for new data structures and comparison. The diagrams show the basic principle. Granularity is broken down to a smaller level, such as the memory resident pages of a task, and the operation is terminated by a context switch, system call, cache flush or similar event. The new areas are compared and, if there is no error, the new blocks become the backups and the system has successfully completed a processing operation. If an error occurs, execution can be repeated using backup data and replacement processors, memory or other hardware. Performance is better than that of file comparison but an error still causes a considerable amount of data to be recalculated. With a file comparison configuration, the repair time is measurable in seconds, while that of a memory comparison is in milliseconds.

An alternative is to run pairs of processors in a lock step, where each operation is checked for errors. This provides the earliest possible error warning but at the highest component cost. Devices like the M68000 family can easily be configured for such designs. Typically, a control signal such as the address strobe is used to start the signal comparison.

The strobe is synchronised from both processors and is used to enable signal comparators which check the address bus, read/write lines, etc. These should be stable when the address strobe is issued. If an error occurs, a bus error can be generated which terminates the cycle and starts exception processing which can investigate the problem. To check data activities, the data strobes can be used in a similar way. There are several considerations however.

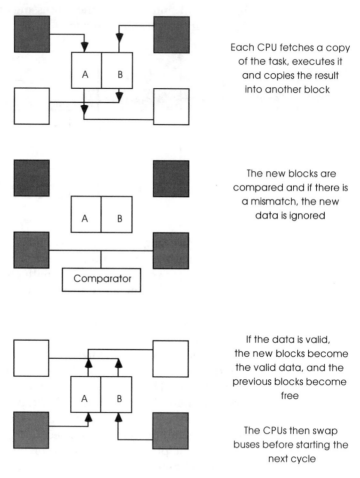

Each CPU fetches a copy of the task, executes it and copies the result into another block

The new blocks are compared and if there is a mismatch, the new data is ignored

If the data is valid, the new blocks become the valid data, and the previous blocks become free

The CPUs then swap buses before starting the next cycle

A memory comparison fault tolerance scheme

The use of stack frames and the data that they contain can cause false errors when their contents are compared. For example, the M68000 processor stack frames contain a mask revision which could cause a data comparison problem. It is important to make sure that processor pairs have the same mask revision. A second problem concerns setting up any internal resources, such as caches and memory management units. The comparison can be disabled while the processor resets and starts up but there is no guarantee that the two processors will be exactly synchronised. A simple solution is to build a common stack frame in external memory and to execute an RTE instruction. This loads back the stack frame and allows the processors to start in a predetermined way. External address translation tables and floating point frames can be loaded to set up other resources. Caches can be preloaded by simple execution of NOPs or simple data moves where explicit clearing or resetting is not available.

The main difficulty concerns hardware comparators and synchronisers. These insert timing delays which can prevent a processor from running as fast as it can. A 20 MHz processor may only be capable of 16 MHz operation when the propagation delays caused by the synchronization logic are added. This effectively places artificial limitations on the performance that can be obtained from such a design.

The M88000 family allows two processor nodes (i.e. an MC88100 and its associated MC88200s) to be run in lock step without any performance impact. Each device can be configured as either a master or a checker. The signals are paralleled up but only the master actually drives the bus. The checker receives the signals and if they correspond with its signal, the processor is allowed to continue. If there is a mismatch, an error signal is asserted and processing is effectively stopped. By bringing the comparison logic onchip, performance degradation is removed, allowing such designs to improve performance through speed upgrades. It must be said that while this feature was perceived as having many benefits, it was a function that fell into disuse and was dropped from the MC88110. The main reason was that providing a CPU check in this way only addresses part of the problem with a fault tolerant design and does not help with memory faults, bus and connector problems and difficulties with peripherals.

Single- and multiple-threaded operating systems

Given a good hardware implementation of a MIMD machine, there is yet another system layer to consider. Most operating systems have been written to support a single processor and therefore need modification to cope with multiple processors. The two methods used are described as single- and multi-threaded, which refers to the way the system handles operating system calls. In a single processor, the supervisor state effectively handles all operating system matters and can be viewed as a separate processor dedicated to that function. The supervisor processor has a single entry point (i.e. the trap instruction) and a single exit point (i.e. the RTE instruction). With a multiprocessor system, there is a choice — either only one processor can act as the supervisor at a time, the single-threaded approach, or multiple processors can, the multi-threaded design.

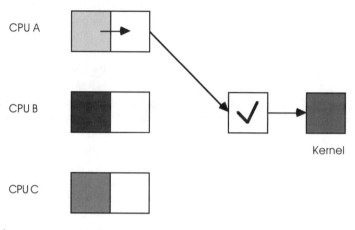

Single-threaded operation – stage 1

With the single-threaded approach, the operating system software is common to all processors and access to the routines and resources is controlled by semaphores. A task running on a processor makes a system call which causes the processor to switch into its supervisor state.

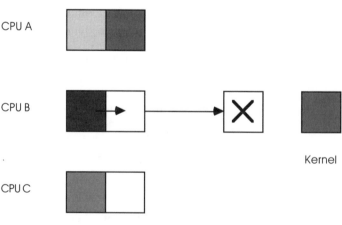

Single-threaded operation – stage 2

It checks to see if any other processor is acting as a supervisor and, if so, waits until the access semaphore is clear. It then sets it and effectively performs a context switch so that its supervisor state is exactly that of the last user. It can now access whatever resource it needs to complete the call. At any one time, only one processor is performing operating system tasks. The second stage shows another processor trying to execute a system call and being blocked by the semaphore.

This type of system is easy to implement. The only real modification is the addition of the semaphore software and associated routines. The bulk of the kernel is still executed by a single processor and needs no modification. The disadvantage of the system is the poor system call performance. There is still only a single processor actually executing the kernel and therefore the system performance in this area is no better off than a single processor system. Applications will run faster, however, and examination of systems which show this type of characteristic are indicative of a single-threaded system.

This configuration is inefficient, considering it is quite common for differing processors and their tasks to need quite separate functions to be performed by the operating system. Obviously, turns must be taken when accessing common resources, such as hard disks, but for other functions it should be possible for the kernel to be split into parts to allow multiprocessing within itself. The multiple-threaded kernel takes exactly this idea. Providing there is no conflict of interest, any processor can perform system functions. The advantage is the speed improvement shown by system intensive applications, but this is offset by the additional software and reworking needed to modify an operating system kernel.

9 Application examples

MC68020 and MC68030 design techniques for high reliability applications

All microprocessors depend on the integrity of the memory subsystem to allow the system to execute reliably. Many early microprocessor families suffered from the execution of illegal op codes, caused by memory corruption, which often put the processor into an indeterminate state — with disastrous results for the external system. To prevent this, parity and/or error detection and correction schemes for memory systems were implemented. Unfortunately, processors were not capable of detecting such memory faults directly — the bus cycle is terminated normally and a non-maskable interrupt generated, allowing software to resolve the situation. This is complex and time consuming. The M68000 family solved these problems by allowing external logic to terminate the bus cycle with an error flag and restart it if required.

Due to the internal microcoding present in M68000 family processors, any attempt to execute an illegal instruction causes an exception to take place. This immediately starts the illegal op code handler software routine which determines what has gone wrong and, more importantly, what to do about it. The processor also identifies incorrect memory addressing, privilege violations, divide by zero and many other errors. Each has its own vector in the vector table and associated software handler, allowing such errors to be controlled, the system repaired and restarted.

This neatly takes care of any obviously wrong memory corruptions but fails to identify corruptions resulting in valid instructions or addresses. To prevent the processor outputting garbage, garbage must be prevented from being input. This requires some external memory checking.

The most common methods use either parity bit generation or error detection and correction. Both of these methods rely on check bits being fetched and compared with the stored data. Parity is only capable of detecting a single bit failure while EDC (error detection and correction) can reconstitute the original data and recover from the corruption. However, such memory protection often cannot signal the error condition until late in the bus cycle or requires extra time to place the correct data on the bus.

Synchronous systems, which have fixed cycle times, incur either additional software overhead in repairing the damage caused by receiving invalid data, or have to run permanently slower to allow time for recovery. Both of these schemes place unacceptable burdens on design.

If invalid data is presented to the processor (due to some bus fault etc.), which is a legal instruction or corrupt data, the software may not recognize the error and start to perform unexpected operations, such as addressing non-existent memory. This is often the first indication that something has gone wrong with software execution. If such a non-existent address is accessed via its asynchronous bus, the processor waits indefinitely for a valid response and appears to hang up. This is easily solved by using a watchdog timer, such as a monostable or clock counter, which is enabled and reset by the address strobe at the start and completion of a bus cycle. Failure to reply with DTACK causes the timer to timeout and assert the BERR line and start bus error or exception processing. This technique not only identifies addressing errors, but can also be used to highlight 'bus hogging' errors in multiple master systems with additional processors and DMA devices.

If the frame building operation for the bus error exception encounters another bus error, the processor assumes that all is lost with its external circuitry and halts. Assertion of the HALT line could fire another timer which attempts to reset the processor and restart the system by asserting RESET and HALT lines. Alternatively, with the processor in a halted state, there is no software resetting of a software timer to reset the processor on expiry. By using investigative software routines within the power up sequence, battery backed-up RAM can be examined and damage repaired or some assessment of data lost etc. made.

The above methods can provide protection from memory corruption caused by hardware failures but do not prevent software overwriting programs or data in error.

Partitioning the system

Many earlier 8 bit microprocessor systems could easily be crashed by bad programming and/or software errors, usually associated with application tasks rather than the operating systems used. The reason was quite simple. There was little or no separation or protection of the controlling software and application tasks running under it. Applications could easily manipulate system tables and stacks and even overwrite kernel areas. Obviously, this was not acceptable for high reliability and is one of the many reasons behind the slower uptake of microprocessors in critical designs.

To solve this deficiency, the M68000 family uses the concept of two different states, with separate and protected resources. The processor normally executes code in the USER state but when an exception occurs, it switches into the SUPERVISOR state, with its own separate stack pointer and access to the processor control registers. This transition can be caused by a processing error, external interrupts or a software interrupt TRAP instruction. Using these two states allows application and operating system software to be separated, preventing rogue applications from corrupting the whole system.

The processor's dual status needs to be repeated in the external design. If the whole of external memory was available to both the USER and SUPERVISOR states, any program would be put at risk from corruption. The memory system must be partitioned so that programs can be protected.

This function is performed by the three function code signals which are driven with the address bus on the start of every memory cycle. The three pins offer eight codes indicating the type of transfer taking place, i.e. program or data, SUPERVISOR or USER, interrupt acknowledge/CPU space cycle. By taking these outputs and including them in the address decode logic, blocks of memory can be reserved for specific use.

Care must be taken with system design if such hardware partitioning is used. Data which is accessed on start-up may be stored in the program boot-up ROMs. While instruction accesses create the correct address and function code information for the decode logic, any data access is inhibited, resulting in a major system problem. It may be impractical to have additional ROMs onboard.

Both the MC68020 and the MC68030 have a solution. The SUPERVISOR has access to two additional registers which allow user-definable function codes to be stored. These can be used with the MOVES instruction to transfer data from one address space to another. When this instruction is executed, the normal function code is replaced by the contents of either the source or destination function code register, depending on which way the data transfer is performed. This allows the correct code to be output, fooling the decode logic and allowing the correct data access.

This ability can also be used to restrict access to I/O devices by mapping them into an undefined address space. Normally, all access to I/O devices is performed in the same way as normal memory access, using the same instructions and addressing modes. If a system breaks down, corrupting its memory, it is possible for the system I/O to be corrupted and left in an indeterminate state. An addressing error can change a parallel port´s status instead of a memory location in a data structure. By placing the I/O addressing in a separate address space, access can only be via the SUPERVISOR, executing the MOVES instruction, with the alternate function code registers set correctly. This reduces the possibility of any corruption and results in such a low level of accidental access that it can be ignored. In addition, any access can only occur via the SUPERVISOR and it can, therefore, be carefully checked.

For many critical applications, additional testing is needed — software requires mathematical verification to provide confidence in its functionality and to establish if there are any design or coding errors. This often places additional restrictions on software design and can prohibit the use of interrupts or exceptions, complicated branching and addressing modes. Indeed, some implementations do not use subroutines, relying solely on inline code and explicit addressing modes. Restricting the instruction set is a simple matter for the compiler. The major problem arises with preventing exceptions.

Consider some safety critical software that has been written and mathematically verified. Such verification inevitably prohibits the use of interrupts and exceptions. Should any occur, software and hardware validity is broken and the system becomes suspect. Such cases should result in the system halting, preventing any erroneous action and possibly allowing a redundant unit to take over. While the majority of exceptions can be either not used or masked, there are some, such as a level 7 interrupt, that cannot be excluded. Although the hardware design can pull the interrupts pins permanently high, an external hardware failure could generate such an interrupt, even though all three pins would need to be asserted simultaneously. Interrupt handlers can be written to return execution to the point prior to the interrupt. If the system has started to malfunction, there is no guarantee that this software will work.

There is a similar problem with the bus error signal. If asserted during a memory cycle, it can cause the bus error exception to be taken. Again, this destroys the mathematical verification and system integrity.

Inhibiting exceptions

Both the MC68020 and the MC68030 have three stack pointer registers available to the programmer. Two are restricted for supervisor use and the third is for the user. The user stack pointer, A7 or USP, is normally used for information and subroutine stacking operations. The supervisor uses either its master stack pointer or interrupt stack pointer, depending on the exception cause and the status of the M bit in the status register. If this bit is clear, all stack operations default to the A7´ stack pointer. If it is set, interrupt stack frames are stored, using the interrupt stack pointer, while other operations use the master pointer. This effectively allows the system to maintain two separate stacks. What is important is that a throwaway stack is built onto the master stack before the complete frame is built using the interrupt stack pointer.

If either of these operations encounters a bus error while building either of these stacks, the processor takes a double bus error and halts. The combination of these two characteristics allows the selective disabling of all exceptions or interrupts only under software control.

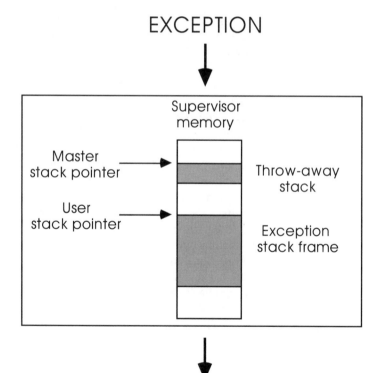

*Normal exception
stack frame building*

A 'no exception' scheme can be implemented using the various stack pointers. On entry to a critical software routine, the stack pointer register is loaded with the address of a non-existent memory location, any stack frame building results in a bus error and halts the processor without any software intervention. If the interrupt stack pointer is so manipulated, only external interrupts effectively result in such a processor halt. If the master stack pointer is used, all exceptions result in a halted processor by virtue of the throwaway stack which is always created on the MSP.

All that is needed is a single instruction at the beginning of the critical routine and another at its completion to implement this concept and increase the overall reliability of the system.

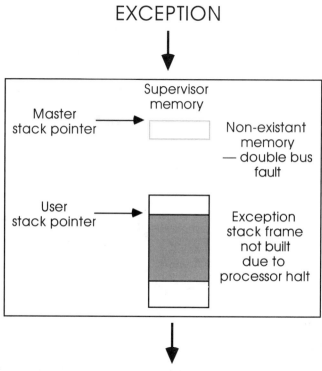

EXCEPTION

PROCESSOR (and system) HALTED

Forcing exceptions to halt the processor

There is another option available which is even more selective. It uses the vector base register (VBR) to relocate the processor vector table to a different memory location. With a new table, individual vector locations can be corrupted so that they cause an address error. This forces the processor to take the appropriate exception which, if it causes an address error, forces the processor to take the double bus fault and halt. This corruption can be caused by making the exception handler pointer stored in the vector table an odd value. Alternatively, the vector and the bus error vector could point to non-existent memory and generate the double bus error this way. This technique can be performed on the original vector table but it uses a lot of overhead to save, modify and restore the vector table. Using the VBR allows one instruction to switch into the protected mode, and another to switch back.

The above technique can be performed by using the software handler to simply execute the HALT instruction. If the memory system fails, the HALT instruction can modified and not executed. The previous method uses the processor hardware and does not rely on external software and its integrity.

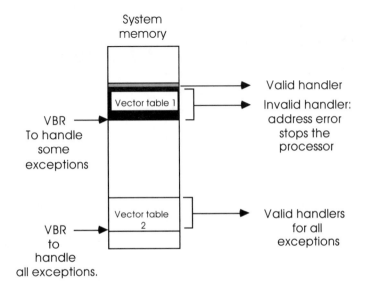

Using the VBR to inhibit exceptions

Implementing the system

The diagram depicts a MC68020-based computer system and its hardware interfaces to external systems. The various memory chips and peripherals are decoded from the full address bus, function code and read/write signals, which are driven at the start of each external bus cycle. To save time, they are decoded and then validated by the address strobe signal AS*. The full address bus must be decoded to ensure that each memory block can only be accessed by addresses within a specific range. If partial decoding is used, each memory block is repeated throughout the entire memory map. This may save on circuitry but does allow access via incorrect addresses. Should an address pointer or an external hardware failure occur, it may access such a multiple mapped block — with disastrous results.

As previously described, it is desirable to limit any I/O access to a specific code sequence using a single privileged instruction to reduce the probability of an error generating a code sequence that might access the critical I/O to virtually zero. This is done by fully decoding the function codes to split the system memory into various blocks — SUPERVISOR, USER, data and program and I/O space. I/O space is defined by using one of the spare undefined codes for critical peripherals. Normal instruction execution cannot generate the right function code to access the I/O space. The address may be generated in error but, without the correct function code, no physical access is performed. Only the supervisor can access such peripherals, executing the MOVES command after the source and destination registers have been loaded with the correct function codes. This limits access to a single instruction requiring several correct register contents, rather than virtually any valid instruction.

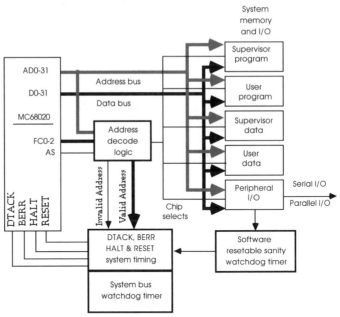

A MC68020 system block diagram

It is important to ensure that the RESET vector is present at location $00000000 and is mapped into the SUPERVISOR data space, so that it can be correctly accessed on power up. A common technique is to remap EPROMs so that for the first few memory cycles, they appear at address $00000000 instead of their normal location. This allows the processor to get its RESET vector, locate the first instructions and jump to the correctly mapped routines in EPROM. The vector table can now be written into RAM at location $00000000. This neatly saves the problem of reconciling a RAM-based vector table with non-volatile storage of the RESET vector.

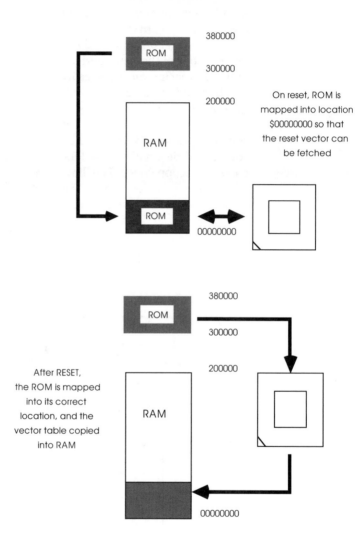

On reset, ROM is mapped into location $00000000 so that the reset vector can be fetched

After RESET, the ROM is mapped into its correct location, and the vector table copied into RAM

A hardware method of providing both a ROMed reset vector and a vector table in RAM

This system can be synthesised in software using the VBR (vector base register). The EPROMs are mapped at $00000000 with RAM located above. Once the EPROM resident software has started, it writes into the VBR, relocates the vector table into RAM and copies its contents from EPROM to RAM. The former solution relocates the EPROM using additional logic; the latter uses a couple of instructions and no additional logic.

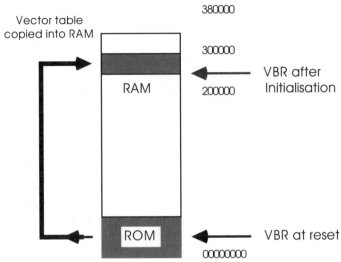

Vector table copied into RAM

380000

300000

RAM

200000

VBR after Initialisation

ROM

VBR at reset

00000000

Using the vector base register to provide a RAM-based vector table

If any data is stored in EPROM, either the EPROM must be mapped into the data space as well or software must transfer all data into RAM using the MOVES instruction during the system set up. This is often neglected and can be a very costly mistake.

Further protection is supplied by a system sanity timer. This requires periodic resetting by software to prevent it from expiring and asserting a combination of system HALT and RESET lines. To prevent accidental access, this timer is mapped into the I/O space with other critical I/O functions. Depending on the required system characteristics, it could either reset the system or halt it.

Bus error handling

The most likely cause of an unexpected exception is the bus error. The initial problem is to determine where the error occurred and then to decide if it is a transitory or permanent fault. By investigating the stack frame built by the processor, the access address, function codes and status of the read/write line, together with the instruction and status registers at the time of the error can be examined. By comparing this information with a system resource map compiled during initialisation, the fault can be diagnosed as an external hardware fault (i.e. an access to an invalid address), or the result of corrupt program or data. If it is hardware, the instruction can be restarted just in case it was a transient problem. With a software fault, all that can really be done is to reset stack areas, note the failure and restart at a known point — i.e. execute the STOP instruction and wait for the next interrupt. If everything is bad, a software reset may be called for. This is achieved by noting the time and nature of the fault in the system parameters memory area, placing the processor in a halted state by executing the HALT instruction and waiting for the external reset watchdog timer to reset the system.

The MC68020 and MC68030 processors provide some additional protection against bus errors generated due to transient errors. If the error occurs during a data fetch, the exception is always taken. If it happens while an instruction is fetched, the exception is only taken if the instruction is executed. The processor prefetch mechanism often fetches instructions which are not executed due to program flow changes, external interrupts, etc. This improves system reliability with respect to transient errors.

Other exceptions

The MC68020 processor will trap other exception conditions and allow their separate handling. Some of them, like divide by zero, misaligned addressing, check bounds violation, etc., are primarily due to hardware faults. All 256 vectors are handled, albeit with a general handler that simply notes the exception and returns. This is essential because any vector can be selected if there is an external hardware fault during a vectored interrupt acknowledgement cycle. For this reason, an autovectored interrupt hardware scheme is used. This limits the choice to the vector for that particular interrupt level and does not use an externally supplied vector to find the exception handler. However, all the vectors are covered in case of a hardware fault.

Upgrading 8 bit systems

The main reason for moving away from an 8 bit processor design to 16 or 32 bits is to provide the next level of performance or functionality within the system. Increased performance either for greater functionality or higher throughput is obvious but many upgrades are done for other reasons. The need to go beyond a 64 Kbyte address range or bank switching to allow more data or program space, or to change from assembler to a high level language to reduce program development and reduce maintenance are often reasons. In some cases, there may only be a need to prepare for a 16/32 bit platform or be able to offer such a solution which pre-empts the move.

Given that an upgrade is needed, there are several ways to proceed. The first is to simply create a 16/32 bit processor design from scratch. This may be the only solution if the new specification is radically different from its predecessor but for many, this is costly, time consuming and effectively scraps the previous design investment.

Comparing a 16 bit MC68000 design with that of a MC6800 or Z80 8 bit system shows that there is remarkable similarity between the systems. Often the main difference is simply the wider memory. A simple transition can be made by using an 8 bit external bus MC68008 processor to allow the memory and peripherals to be used with an internal 16/32 bit processor architecture. The memory array, MC68901 Multi-Function Peripheral and EPROM are byte wide. 16 and 32 bit accesses are performed automatically as a sequence of byte accesses. The processor can provide up to 4 Mbytes of direct addressing. It achieves about 60% of the throughput of the full 16 bit MC68000. The figure is greater than half because byte data fetches take the same time with the MC68000 and MC68008 but instruction and word fetches take half the time.

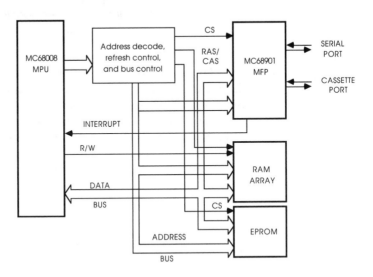

An MC68008 basic system

This approach has other advantages within a bus-based system, in that it allows a simple processor board replacement.

The M68000 asynchronous bus interface has simpler requirements than many 8 bit processors. It only needs a single clock, which can easily be generated from a combination of a crystal and a 7404 inverter chip. This provides the system clock from which the control signals are synchronized, i.e. inputs are recognized, providing a small setup time prior to the clock edge is met and outputs appear within a fixed time from the clock edge. The non-multiplexed data and address buses remove any need for demultiplexers or latches.

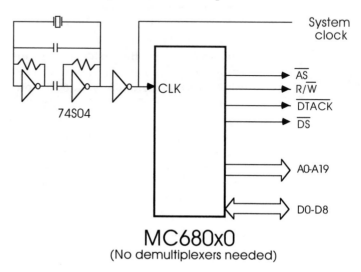

MC680x0
(No demultiplexers needed)

The MC680x0 clock circuit

The asynchronous bus allows any speed peripheral to talk to the processor and this greatly simplifies interfacing other peripherals to it. For existing M6800 peripherals a M6800 bus, complete with E clock, is supplied for easy upgrading. The E clock is generated from the processor clock by dividing it by 10. For 8 and 10 MHz devices, M68Axx parts are suitable and for higher processor speeds, M68Bxx chips are needed.

With an asynchronous interface, it is possible to force the M68008 to wait indefinitely if the cycle is not terminated either by a DTACK* or a BERR* reply. The diagram shows a circuit which can generate DTACK* with differing timings and a BERR* signal if the access is to invalid memory. The counter simply clocks the system clock once it has been enabled by the address strobe, AS*. This only happens with a start of a valid bus cycle. The counted outputs can then be gated with valid address signals to provide the appropriate delay. Using a jumper field allows the user to modify such timings if faster memory, peripherals or processor are added.

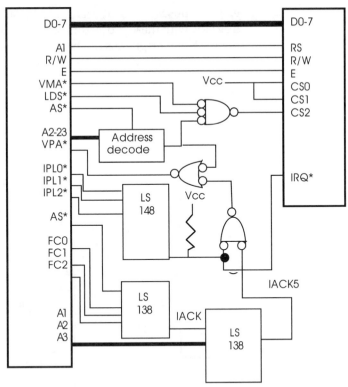

E clock generated from system clock divided by 10.
For 8,10 MHz CPUs use 68Axx parts
For 12, 16 MHz CPUs use 68Bxx parts

*Interfacing an
MC68000 to M6800
peripherals*

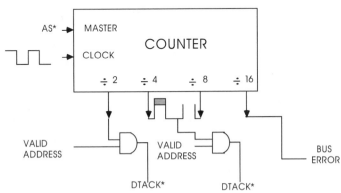

*Generating bus error
and DTACK*

For systems that need a full 16 or 32 bit pathway, the
MC68008 is not suitable and either a 16 or 32 bit member of the
family must be used. The MC68000 and MC68010 have 16 bit

external buses and still retain the M6800 interface bus. To upgrade with these processors requires a 16 bit memory path but the 8 bit peripherals can be retained. Many of the M68000 peripherals still have an 8 bit data bus because many of their I/O functions are still based around byte data sizes.

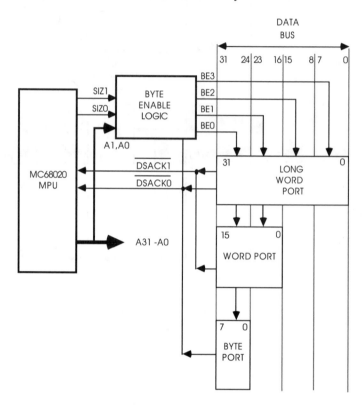

DSACK1	DSACK0	MEANING
HI	HI	Insert wait state
HI	LO	Complete cycle, port size = 8 bits
LO	HI	Complete cycle, port size = 16 bits
LO	LO	Complete cycle, port size = 32 bits

The MC68020 dynamic bus sizing

A variation on this can be provided by the MC68HC001. This is an HCMOS version of the M68000 running at 16 MHz and has the ability to be configured as either a 8 or 16 bit external bus on reset by using an additional 'mode' pin. The device provides a higher speed, lower power consumption upgrade for the MC68008 and allows dual designs to be built. Such a board would be designed for a full 16 bit memory interface and 8 bit peripherals. For lower cost and lower performance systems, the extra components needed for the extra 8 bits would not be inserted into the board and the MC68HC001 jumpered for 8 bit mode. To achieve higher performance, extra memory and associated buffers etc. are inserted and the processor mode changed to 16 bit. This gives a lot of flexibility for system upgrading without redesigning.

For low cost 32 bit systems, the MC68020 and MC68030 can offer good performance with the ability to use 1, 16 or 32 bit wide memory ports without prior knowledge of the port size. This gives the option of mixing and matching 8, 16 and 32 bit peripherals and buses as necessary.

Dynamic bus sizing works by the processor always assuming that the memory port is 32 bit wide. The memory port accepts whatever width of data it can, tells the processor how much data has been taken and how much is left to transfer via a combination of DSACK* signals. The processor simply moves the data up on the bus and performs another bus cycle. The cycle repeats until all the data has been transferred.

Transparent update techniques for digital filters using the DSP56000

One of the most frequently used DSP algorithms is the finite impulse response filter (FIR), which is used to perform filtering for tone detection within telecomms applications, tone filters and also for special effects in digital audio and pitch detection within speech recognition systems. The FIR is commonly used because it is non-recursive and therefore inherently stable. It is also easy to adapt through manipulation of its coefficients tables.

Coefficients are the analogue equivalent of the resistor–capacitor networks used in analogue filters. These coefficients need to be changed to adapt the filter´s response or if the algorithm needs modification. In either case, a controlling microprocessor, like an M68000 or an MC68HC11, can calculate the new values and transfer the data to the signal processor. The challenge is to find the most efficient method of transferring the data.

Method 1

A simple way involves creating a shared memory system between the host processor and the DSP56000. The host processor calculates the new data in its own memory, arbitrates the DSP56000 off the bus, transfers the data, signals the availability of new data and lets the DSP56000 back on to the bus. The DSP56000 checks the semaphore and switches algorithm or coefficients.

This solution is viable, but it has certain disadvantages. It is extremely hardware intensive, needing a bus arbiter, dual port or shared memory and controlling logic. The overhead of checking the semaphore is high and occurs irregularly. The DSP56000 either slows down or stops execution during the external fetches or when the host processor is accessing the bus.

Method 2

The DSP56000 has a host interface port which can be used by the host processor to pass data into the chip. The interface simply looks like a memory mapped peripheral to the host processor.

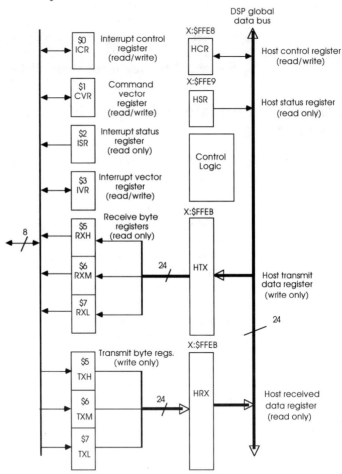

The DSP56000 host interface port

The 8 bit port on the left is accessible by the host while the transmit and receive buffers, together with the control and status registers, appear as part of the DSP56000 global data bus. The host processor can force the DSP56000 to execute a fast, two-instruction interrupt routine which can be used to transfer data for example.

With this interface, the block transfer can be made directly into the DSP56000, reducing the necessary hardware and removing any arbitration delays. However, this improvement has some problems. Consider the block transfer time. For a 100 tap filter with 24 bit coefficients, 300 bytes need to be moved. Using a 16 MHz M68000 with a 62.5 ns clock time and three MOVE.B instructions with incremental indirect addressing, takes 275 microseconds. This may not appear to be a very large overhead but, coupled with the synchronization overhead, it is sufficient to reduce the effective sampling rate to 3.6 kHz.

Method 3

The best method also utilises the host interface but uses the host command vectors and fast interrupt to effectively cycle steal from the DSP56000. Each 24 bit coefficient is written into the host interface and a fast interrupt initiated to force the DSP56000 to execute one of 16 host command vectors.

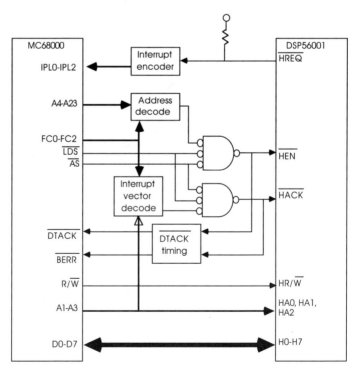

A MC68000/DSP56000
multiprocessor
configuration

These routines can transfer the data into internal memory and switch coefficient tables as necessary. The block transfer of data from the M68000 to the DSP56000 is done in parallel with the filter execution and is distributed across a larger time period. This effectively places no overhead on signal processing.

At 44.1 kHz and 1 transfer per cycle, throughput is only reduced by 0.9%. In addition, there is no impact on I/O service or sampling rates, minimal hardware is needed and synchronization is controlled by the host.

Motor and servo control

An increasing need for microprocessor designs is to interface with the real world to drive motors and other mechanical equipment. The hard disk controller is a good example of such needs and shows how a processor can be used to implement the necessary control and servo algorithms.

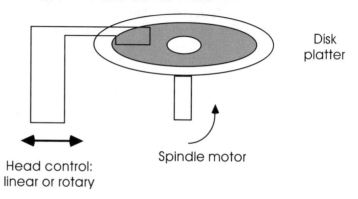

Disk platter

Head control: linear or rotary

Spindle motor

Both controls usually three-phase brushless DC motors

Hard disk mechanics

Hard disk mechanics usually consist of two main functions, as shown in the diagram. The disk platter speed must be kept constant and within certain strict tolerances to maintain data integrity. The reason for this is simple — the data is recorded by magnetising the metal oxide on the surface of the disk platter to form concentric rings of data called tracks. Each track is divided into sectors, with each sector storing a certain number of bytes and error correction codes. The speed of the platter and the resolution of the data head determine the recording density. If the speed changes, the recording density will differ from the time of writing the data and reading the data. These differences can be so great that the original data cannot be read correctly. In such cases, data is effectively, lost causing much grief to the unfortunate user! This change in speed is the reason IBM format 3.5" disks are incompatible with Apple Macintosh disks. While the actual floppy disks are identical, the IBM PC format is based on a constant rotational speed; the Apple format changes the rotational speed, depending on which track is being accessed. In both cases, it is essential that the rotational speed during recording and reading is the same.

The head actuator also requires sophisticated control. The tracks are extremely narrow and yet each track must be repeatedly located with great accuracy many hundreds of thousands of times throughout the hard disk's life. In addition, the seek time taken to locate a track must take the shortest time possible. With a hard disk system, it is the seek time that plays a large part in determining the data access time and the user-perceived response time.

System needs

The next diagram shows a typical system and how the electronics falls into two main areas. The drive controller takes the data via an interface, like ST506 or SCSI (small computer systems interface), and the appropriate location data (e.g. track and sector number), passes the data to the read/write electronics and moves the head into the appropriate position.

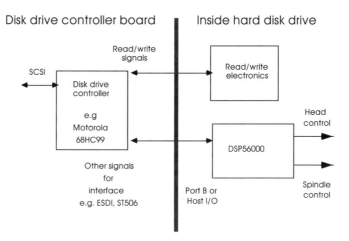

Typical electronics

Read/write electronics usually consist of a low noise analogue amplifier which directly drives the recording head. Digital data is converted by the controller to a analogue serial data stream before being passed to the read/write heads, in a method similar to that used within MODEMs.

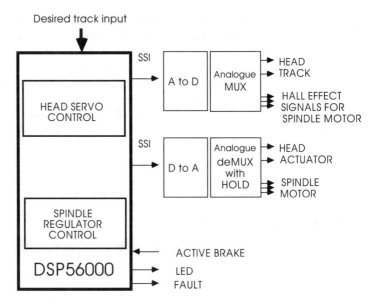

Desired track input

DSP56000 functions

The spindle and head control uses standard servo designs where error signals are constantly fed back to the controller so it can detect and compensate for any deviations. This type of function can be performed by any microprocessor with timer, analogue and digital outputs. The servo response time and positional accuracy are effectively determined by the computational powers of the processor. For many lower performance designs, higher performance microcontrollers, like the MC68HC11, are suitable. As performance increases, more and more arithmetic calculations are needed which can only be met by the faster M68020/MC68030, M88000 or DSP families. Considering the space limitation and cost restrictions, the DSP56000 offers a compact, powerful solution. The diagram shows how the DSP56000 can be configured. The main control signals require conversion from the analogue to the digital domain and this is performed by the A to D and D to A converters located on the Synchronous Serial Interface (SSI).

Spindle control

The diagram shows the basic algorithm used to control the rotational speed. External Hall effect position sensors measure the rotational speed of the spindle and generate time-based event data. This is taken into the processor and compared with a standard velocity event generator. The difference in the timing between events can be measured and passed to a compensator which calculates the required compensation and generates the appropriate signals to control the spindle motors and restore or maintain the correct velocity. These functions are are performed in software, using simple control techniques.

Control signal generation is the most difficult. The speed and granularity of this data determines the servo response but the generation of sine wave data from digital sources is difficult. There is a further complication involving the constantly changing phase relationships between the three outputs needed to control the three spindle motors. The DSP56000 has an elegant solution. It has a sine wave table digitally encoded within its data ROM and the modulo addressing can be used to cycle through this table without end looping overhead. By using three pointers, differing phase relationships can be created by varying the address space between them, as shown.

Head control

The head positioning algorithm is very similar to that used to control the spindle. The required track position is given to the DSP56001 and this is used for comparison with the current track position and generating the required control signals. If the head is at track 12 and the command to move to track 67 is given, this causes the DSP56000 to start the head actuators moving in the given direction. The movement data is based on current velocity and direction and uses a commutation table and the embedded sine wave table to generate the correct control signals. A notch filter is included to remove effects due to mechanical resonance etc. There are two main considerations. The first is to detect each track and count it and the second involves adjusting the actuator velocity to compensate for mechanical inertia.

In the same way that a vehicle cannot stop dead and needs a stopping distance, the inertia within the actuator requires the stop command to be issued before it actually is positioned over the track. This requires velocity compensation so that the head can be moved quickly over a large number of tracks but can also be moved slowly when fine control is needed.

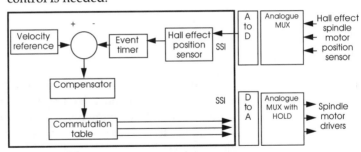

Spindle control

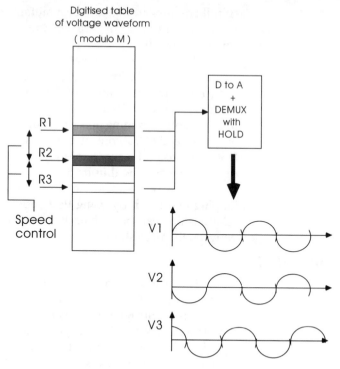

*Generating AC
voltage waveforms*

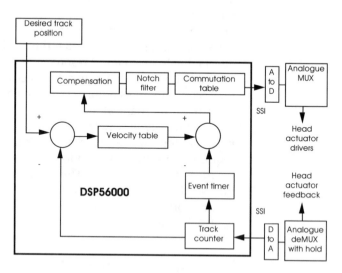

*The positioning
algorithm*

The technique used with the DSP56000 can provide
both the track count and velocity information. The SCI timer
is programmed to provide a succession of timer interrupts.
This allows the velocity between each track position to be
calculated, as shown. Each track position generates an IRQA

interrupt and the number of timer interrupts give the velocity between tracks. The lower the number of SCI timer events, the faster the velocity. Each timer interrupt generates a fast interrupt which, in turn, causes a pointer to be incremented.

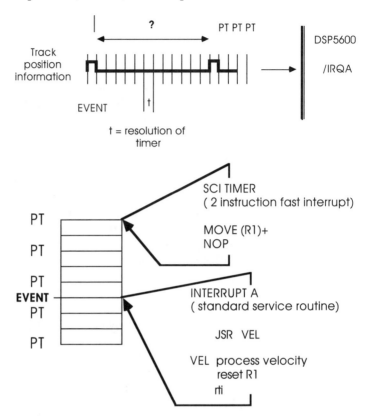

Using the SCI timer

When a track position interrupt occurs, the track counter is increased and the velocity processed using the pointer value. Finally, the pointer is reset before restarting the sequence.

Determining track position

The main difficulty with this type of system is providing the track position information. A low cost solution is to use one of the disk platters to store only track positioning information and keep this as the reference for the other heads within the system. This reduces the drive storage capacity but is relatively simple to implement. This method is sensitive to temperature changes, external shocks and vibrations and component ageing. In addition the positioning of the servo head may not necessarily correspond with correct data head positioning.

An alternative and more costly method is to embed the servo track positioning information within each track. This frees up the servo platter for data storage and increases disk capacity. However, the position control is far greater and requires faster and more accurate responses from the controller. This increases the processing power needed and, with the control functions using digital processing techniques, a DSP chip is almost mandatory. With such a system, track densities of greater than 1,200 per inch are obtainable.

The DSP56000 is ideal for such applications because of its 24 bit precision for the control algorithms, its high parallelism and processing throughput.

Improved SRAM interfaces

When asked to predict how static RAMs will develop in the future, the standard answer is frequently based around shorter access times and higher densities, coupled with the rider that this will keep track with developments in processor technology. It is assumed that they will be little change in the basic SRAM interface and that this is sufficient to meet future requirements. While the advent of faster memories is always welcome, the challenge facing designers today is how to take the traditional interface using non-multiplexed address and data buses controlled by chip select logic and applying it to today's microprocessors which no longer use the multiple cycle memory access.

The burst interface, which is used on processors from Motorola, Intel, AMD, MIPs and many other manufacturers gains its performance by fetching data from memory in bursts from a line of sequential locations. It makes use of a burst fill technique where the processor will access typically four words in succession, enabling a complete cache line to be fetched or written out to memory.

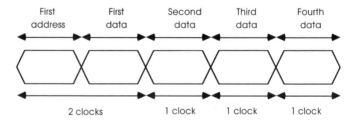

Burst fill interface

The improved speed is obtained by taking advantage of page mode or static column memory. These type of memories offer faster access times — single cycle in many cases — after the initial access is made. The advantage is a reduction in clock cycles needed to fetch the same amount of data. To fetch four words with a three clock memory cycle takes 12 clocks. Fetching the same amount of data using a 2-1-1-1 burst (two clocks for the first access, single cycle for the remainder) takes only five clocks.

Burst fill offers advantages of faster and more efficient memory accesses, but there are some fundamental changes in its operation when compared with single-access buses:

*   The address is only supplied for the first access in a burst and not for the remaining accesses. External logic is required to generate the additional addresses for the SRAM interface.
*   The timing for each data access in the burst sequence is unequal: typical clock timings are 2-1-1-1 where two clocks are taken for the first access, but subsequent accesses are single cycle.
*   The subsequent single-cycle accesses compress address generation, setup and hold and data access into a single cycle, which can cause complications generating write pulses to write data into the SRAM.

These characteristics lead to conflicting criteria within the interface: during a read cycle, the address generation logic needs to change the address to meet setup and hold times for the next access, while the current cycle requires the address to remain constant during its read access. With a write cycle, the need to change the address for the next cycle conflicts with the write pulse and constant address required for success.

The MC68040 burst interface

The MC68040 burst interface shows how these conflicts arise and their solution. It operates on a 2-1-1-1 basis, where two clock periods are allocated for the first access, and the remaining accesses are performed each in a single cycle. The first function the interface must perform, is to generate the toggled A2 and A3 addresses from the first address put out by the MC68040.

This involves creating a modulo 4 counter where the addresses will increment and wrap around. The MC68040 uses the burst access to fetch four long words for the internal cache line. It will start anywhere in the line so that the first data that is accessed can be passed to the processor while the rest of the data is fetched in parallel. This improves performance by fetching the immediate data first, but it does complicate the address generation logic — a standard 2 bit counter is not applicable. A typical circuit is shown on the next page.

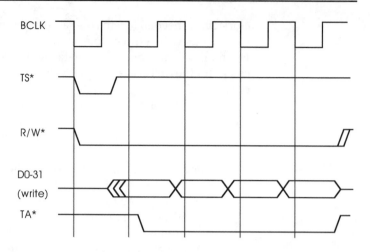

MC68040 burst
interface

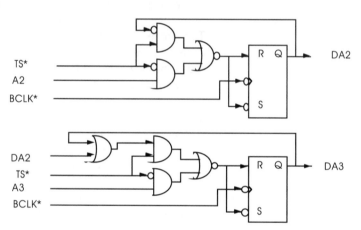

Modulo 4 counter
(based on a design
by John Hansen,
Motorola Austin)

Given the generated addresses, the hardest task for the
interface is to create the write pulse needed to write data to the
FSRAMs. The first hurdle is to ensure that the write pulse
commences after the addresses have been generated. The
easiest way of doing this is to use the two phases of the BCLK*
to divide the timing into two halves. During the first part, the
address is latched by the rising edge of BCLK*. Latching DA2
and DA3 holds the address valid while allowing the modulo
4 counter to propagate the next value through. The falling
edge of BCL* is then used to gate the read/write signal to
create a write pulse. The write pulse is removed before the
next address is latched. This guarantees that the write pulse
will be generated after the address has become valid. This
circuit neatly solves the competing criteria of bringing the

write pulse high before the address can be changed and the need to change the address as early as possible. The logic and timing is shown in the two diagrams

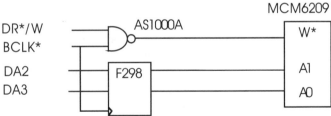

Latching the address and gating W*

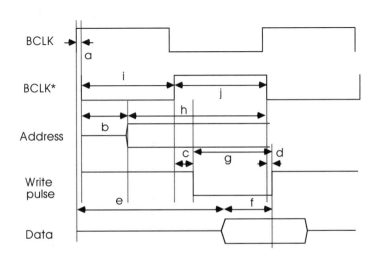

Timing	Description
a	Clock skew between BCLK and its inverted signal BCLK*
b	Delay between BCLK* and valid address — determined by latch delay
c	Gate delay in generating Write pulse from rising BCLK* edge
d	Gate delay in terminating Write pulse from falling BCLK* edge
e	Time from rising edge of BCLK to valid data from MC68040
f	Data set up time for write referenced from = i+j-e+a
g	Write pulse width = j-c+d
h	Valid address i.e memory access time.
i,j	Cycle times for BCLK and BCLK*

Write pulse timings

The diagram shows the timing and the values for the write pulse, t_{WLWH}, write data setup time, t_{DVWH} and the overall access time t_{AVAV}. For both 25 and 33 MHz speeds, the access time is always greater than 20 ns and therefore 20 ns FSRAM would be sufficient. The difficulty comes in meeting the write pulse and data setup times. At 25 MHz, the maximum write pulse is 17 ns and the data setup is 9 ns. Many 20 ns FSRAMs specify the minimum write pulse width with the same value as the overall access time. As a result 20 ns access time parts would not meet this specification. The data set up is also longer and it is likely that 15 ns or faster parts would have to be used. At 33 MHz, the problem is worse.

	Clock Skew (a)	F298 Delay (b)	AS1000 delay (c)	AS1000 delay (d)	Spec 18(040) (e)	t_{DVWH} (f)	t_{WLWH} (g)	t_{AVAV} (h)	BCLK (i)	BCLK* (j)
25 MHz timings										
Small Buffers	0	9.5	4	1	32	9	17	31.5	20	20
Big Buffers	0	9.5	4	1	23	18	17	31.5	20	20
F00+Small Buffers	0	9.5	5	2.4	32	10.4	17.4	32.9	20	20
F00+Big Buffers	0	9.5	5	2.4	23	19.4	17.4	32.9	20	20
Small Buffers	1.5	9.5	4	1	32	10.5	17	31.5	20	20
Big Buffers	1.5	9.5	4	1	23	19.5	17	31.5	20	20
F00+Small Buffers	1.5	9.5	5	2.4	32	11.9	17.4	32.9	20	20
F00+Big Buffers	1.5	9.5	5	2.4	23	20.9	17.4	32.9	20	20
33 MHz timings										
Big Buffers	0	9.5	4	1	20	11	12	21.5	15	15
F00+Big Buffers	0	9.5	5	2.4	20	12.4	12.4	22.9	15	15
Big Buffers	1.5	9.5	4	1	20	12.5	12	21.5	15	15
F00+Big Buffers	1.5	9.5	5	2.4	20	13.9	12.4	22.9	15	15

MC68040 timings for small and large buffers

Meeting the interface needs

For a designer implementing such a system there are four methods of improving the SRAM interface and specification to meet the timing criteria:

*   Use faster memory
*   Use synchronous memory with onchip latches to reduce gate delays
*   Choose parts with short write pulse requirements and data setup times
*   Integrate address logic onchip to remove the delays and give more time.

While faster and faster memories are becoming available, they are more expensive, and memory speeds are now becoming limited by on and offchip buffer delays rather than the cell access times. The latter three methods depend on semiconductor manufacturers recognising the designer's difficulties and providing static RAMs which interface better with today's high performance processors.

The Motorola QuickRAM™ Fast Static RAMS have very short write pulse and data set up requirements. At 25 ns, the output enable versions need only a 15 ns write pulse and a 10 ns data set up. The standard parts at 20 ns would work in this design as well. In both cases, these slower parts are considerably cheaper than faster alternatives. Even at 33 MHz, using memories with output enables, it is possible (just) to meet the timings with 20 ns parts. By replacing the AS1000 gate with a F00 gate and allowing the timings to skew slightly to gain a couple of nanoseconds, a bit more timing margin can be obtained.

This approach is beneficially for many high speed processors, but it is not a complete solution for the burst interfaces. They still need external logic to generate the cyclical addresses from the presented address at the beginning of the burst memory access. This increases the design complexity and forces the use of faster memories than normally necessary simply to cope with the propagation delays. The obvious step is to add this logic to the latches and registers of a synchronous memory to create a protocol specific memory that supports certain bus protocols. The first two members of Motorola's Protocol specific products are the MCM62940 and MCM62486 32K x 9 fast static RAMs. They are, as their part numbering suggests, designed to support the MC68040 and the Intel 80486 bus burst protocols. These parts offer access times of 15 and 20 ns.

The MCM62940 has an onchip burst counter that exactly matches the MC68040 modulo 4 burst sequence. This supports the wrap around line fills that are used by the MC68040 to fetch the data that the processor needs first and immediately passing it through, while the bus interface carrys on with the line fill. This removes some three to four gate propagation delays caused by the external logic. The address and other control data can be stored either by using the asynchronous or synchronous signals from the MC68040 depending on the design and its needs. A late write abort is supported which is useful in cache designs where cache writes can be aborted later in the cycle than normally expected, thus giving more time to decide whether the access should result in a cache hit or be delayed while stale data is copied back to the main system memory.

The MCM62486 has an onchip burst counter that exactly matches the Intel 80486 burst sequence, again removing external logic and time delays and allowing the memory to respond to the processor without the need for the wait state normally inserted at the cycle start. The chip logic automatically inserts a delay to conform with the processor timing without necessarily inserting a wait state, thus improving performance.

In addition, it can switch from read to write mode while maintaining the address and count if a cache read miss occurs, allowing cache updating without restarting the whole cycle.

Conclusions

While faster memories will always be needed to provide the ultimate performance for many processor designs, there are many cases where faster access times are simply needed to overcome limitations within the SRAM and processor interface. Better specifications and changing the bus interface to improve support for a bus protocol are more elegant solutions which do not rely on faster access times to achieve the same end goal. It is an approach that has many merits.

10   Semiconductor technology

The last decade has seen some unparalleled advances in the semiconductor industry. Memory densities have increased by over a thousand times, the simple 8 bit microprocessors of the late 1970s have matured into 1 million transistor plus processors like the MC68040, current handling and switching speeds have increased from a few milliamps to tens of amps and from a few kilohertz to over 200 MHz.

Silicon technology

Integrated circuits are fabricated by taking monocrystalline wafers of silicon and diffusing impurities into them under controlled conditions to form semiconducting structures within the material. These are then connected together using aluminium metalisation or conductive polysilicon tracks.

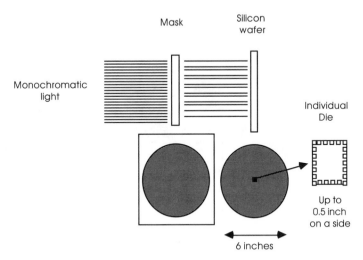

The basic photo-lithography process for integrated circuits

The tracks and diffusions are selectively performed using a photoresist coating, exposing it to a patterned mask, developing the resist to remove the unexposed resist and thus the silicon. Diffusions are achieved using heating within a gaseous atmosphere or by direct ion implantation.

These processes are carried out on a very small and continually decreasing scale. The metalisation tracks used in devices of 10 to 15 years ago were about 10–20 microns in width. This may be thinner than a human hair but is many times larger than the submicron widths appearing today.

Semiconductor structures are formed around P and N types of diffusions, caused by a lack or excess of electrons within the crystal lattice. With these basic blocks, capacitors, transistors, diodes and resistors can be fabricated and connected in a similar way to which discrete components are utilised on a printed circuit board. There are two types of transistor that can be made in this way.

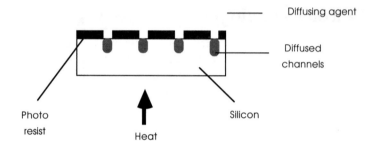

Diffusing channels into silicon wafers

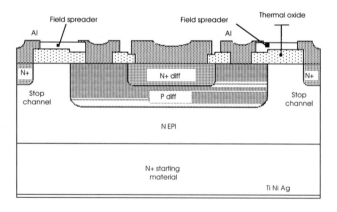

Epitaxial bipolar transistor

The first is the bipolar transistor whose structure in shown in the cross-section. It consists of three semiconductor diffusions to from a N-P-N or a P-N-P sandwich. The central connection is the base and the emitter and collector are the outside connections. In the example shown, the P Diff region is the base, the N+ Diff region the collector and the emitter formed from the N EPI material below it. Bipolar transistors have very fast switching speeds of many gigahertz and good current handling but their construction is large and their power consumption high.

The MOS transistor is small, with reasonable switching capability but with reduced current capability. It works via a field effect which creates a channel below a very thin layer of dielectric, i.e. silicon oxide, when a voltage is applied above it. This gate voltage can be supplied by a metalisation track, in the case of metal gate CMOS, and by polysilicon in most other cases. The MOS transistor forms the basis of most modern digital electronics. For very high speeds, bipolar logic called emitter coupled logic (ECL) is used — but with the penalty of smaller densities and greatly increased power consumption.

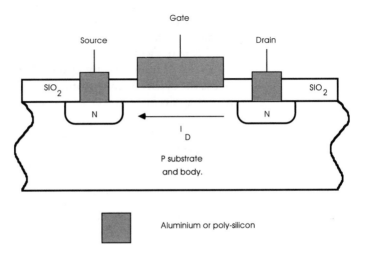

A conventional MOS transistor

CMOS and bipolar technology

The fuel for the processors which have appeared over the recent years is the available semiconductor technology and the transistor density and switching speeds that it can offer. The technology has been based around silicon as the main semiconductor and although more esoteric compounds like gallium arsenide continue to challenge it in raising switching speeds, silicon technology will continue to be the mainstay of the industry. Research teams have developed bipolar silicon transistors with switching speeds of up to 26 GHz. Similar PNP devices have also been reported. Current bipolar technologies will get faster and faster as the transistor size is shrunk. This will increase densities and allow many VLSI devices that currently use CMOS technology to be implemented in bipolar technology, enabling designs like the M88000 RISC family to be implemented in ECL logic and allowing a dramatic increase in processor clock speeds. Although predicted or announced to be the technology for very fast processors in the early 1990s, ECL versions of CMOS processors have never really been successful within the market due to the problems of the heat dissipation and the competion from very fast CMOS processors. As a result, many of the planned ECL processors never reached the market place.

Over the last ten years, there has been a dramatic swing towards CMOS as the technology of choice for many designs. Its reduced power consumption is essential to maintain acceptably low power dissipation as the number of active transistors and speeds increase (CMOS devices use more power the faster they run). It is likely that this swing will continue. The roadmap for ASIC CMOS technology shows today's 100,000 gate devices being dwarfed by the multimillion gate chips that will undoubtedly appear. The reason for this increase is the ever reducing geometries which allow increased densities and reduce propagation delays.

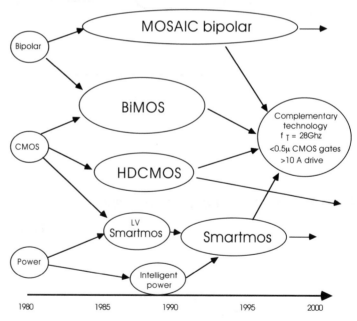

*Motorola's technology
roadmap for the 1990s*

The future of CMOS technology on its own to provide the highest speed processors and memory does appear to be limited. As speeds increase, the required current drive increases to the point where a combination of CMOS logic and bipolar drivers becomes the logical and most effective solution. CMOS is used to provide the logical functions while bipolar outputs are used to provide the high drive. Combination technology, such as BiCMOS, is now starting to appear in memories and will undoubtedly take over as the preferred technology for these types of parts.

This combining of technologies is Motorola´s aim by the mid to late 1990s. The three core technologies of bipolar, CMOS and power devices will be developed and combined to provide various combinations. Bipolar will continue to be developed through the MOSAIC (MOtorola Self Aligned Integrated Circuit) program and CMOS through the HDCMOS series. These technologies will provide a core of complimentary processes which will allow VLSI designs to combine very fast switching speeds of up to 28 GHz, with current drives of 10 A and high gate density.

This combination allows the integration of digital technology with analogue interfaces, will drive more electronics into the controller world and greatly reduce the need for wholly analogue solutions. As the cost of going digital decreases, the advantages of no component ageing or drift, finer resolution, etc. offered by digital techniques may prove irresistible. The trends seen in automotive design over the last few years, where mechanical ignition, braking systems and even light switches have been replaced by electronics, are an indicator of what is going to happen.

Fabrication technology

The key to generating the silicon is fabrication technology. The main issue in the 1990s is not whether visible light will be replaced by electron beam or X-rays, but when. Light plays an integral part in building any semiconductor. It is responsible for transferring the patterns from the mask to the silicon surface to allow the selective diffusion and etching necessary to build a transistor, a logic gate or a processor. The resolution of this process determines the size of the transistors that can be built. Today´s technology is still predominantly light based and is rapidly reaching its limits. The problem concerns the wavelength of light and, therefore, the resolution of the objects it can expose. As an object´s size decreases and approaches that of the wavelength of light, it becomes harder to differentiate it from the background and project it to make the mask or project its pattern onto silicon.

The solution is to use shorter wavelengths. This requires the use of electron beams or X-rays to replace visible light. Electron beams are currently used to fabricate the masks and in direct etching techniques but, to achieve line widths of

0.25 micron and below, it is likely that X-rays will be needed. This work is already underway. Motorola and IBM are co-funding a $1 billion X-ray lithography research programme to develop the appropriate technology.

An interesting implication of this trend will be the exponential increase in capital investment to finance new silicon fabrication plants. It is estimated that a sub micron fab facility costs about $500 million today. The next generation facility, to cope with 0.25 micron geometries and below, is estimated to cost over $1 billion. For many of today´s smaller silicon companies, this level of investment is far greater than their turnover or capitalisation and it is likely that between now and the next century many will either merge, be taken over, or simply disappear. This trend has already started. Mostek and Fairchild are a couple of famous names that suffered this fate in the 1980s.

Packaging

The trend in packaging is quite simple. Smaller packages, larger numbers of pins and smaller pin separation — everything that gives board layout engineers nightmares! Analysis of the last ten years has shown a move away from the dual in line packages, with about a 64 pin limit, to pin grid array packages of about 200 pins, surface mount packages of 200–300 pins etc. The differing types of ASIC package options which are currently available give a good indicator of the standards to be seen in the next few years. Currently, Motorola is offering pin counts up to about 500, using a variety of packages ranging from pin grid array, quad flat pack through to TAB bonding.

With TAB bonding, the die is inverted and bonded directly to the frame fingers. These fingers are brought outside the package to connect to the outside world. The trend is of decreasing lead pitch.

Packaging sizes will decrease down to chip level. This is not new, as hybrids have been around for many years. Here, several dice are mounted on a substrate and bonded together to form a complete unit in its own right. The flip chip technique, where the chip is inverted and directly bonded to the substrate, will provide the highest packing density. Motorola has developed a technique which takes this a step further. A silicon substrate is used with the appropriate diffused connections and holes are then etched in the substrate to accept individual die. In this way, it is possible to achieve a customised 'wafer scale integration', knowing the dice are all functional. The use of silicon dice and substrate prevents many of the thermal mis-matching problems.

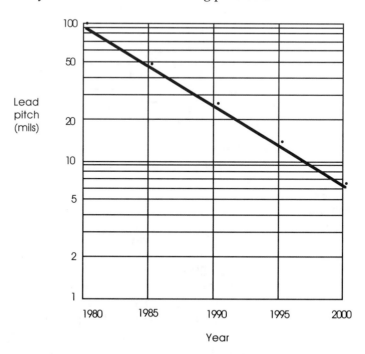

Trends in lead spacing

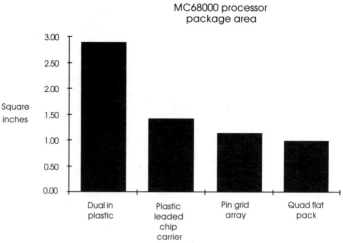

Chip versus package areas for the MC68000

With surface mount techniques becoming more and more common, the package options are bound to increase. Currently, plastic leaded chip carriers (PLCC), ceramic leaded chip carriers (CLCC) and plastic and ceramic quad flat packs (PQFP, CQFP) are the favoured packages offering high pin densities and low cost. Versions of these packages will appear with even smaller leads and higher densities.

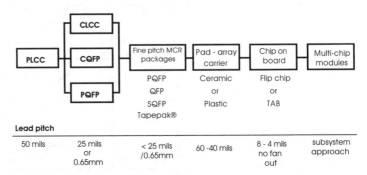

Surface mount technology progression

The pad array carrier (PAC) is another package type which may become extremely widespread. It is typically about 10% bigger than the die and consists of a ceramic body with a matrix of solder bumps arranged on a 50 thousand spacing. The MC88100 and MC88200 are packaged in this form for the Hypermodule to pack up to 4 x MC88100 processors and 8 x MC88200 cache memory management chips on to a single board 8" x 3". This technology gives extremely high packing density and allows science fiction dreams to become reality. The Motorola micro-ETACS cellular phone, which looks like a Star Trek communicator, and the wrist watch pager are proof that this type of technology is not restricted to niche products but is rapidly entering the consumer market, where size is a premium.

These new generations will need advances in board design, surface mount techniques and testing before this state of the art becomes routine. How do you test a PAC device on a board when none of the connections are exposed? The answer is to X-ray it or design the board to provide test points. How do you then rework the board? Solving this sort of problem will require additional help in increasing the testability of a design. JTAG testing, where test signals can be circulated around the pins of a device by putting the device into a special test mode, has just started to appear. The MC68040 has this facility. By the end of the century, probably every VLSI chip will have this or a similar facility and as much attention will be paid to testing systems as designing them.

One aspect that will affect all electronic engineers is the need by law to minimise radio frequency interference (RFI). With chips becoming faster and offering greater current drive, more care will be needed on layout to meet the new EEC and FDI specifications concerning such emissions. This will involve the large-scale adoption of transmission line designs for all active signals and special care will be needed during layout to prevent the physical design from increasing such problems. Fortunately, the techniques used to reduce RFI are good design practices and, apart from a forced renewed interest in electromagnetic design theory, this should pose no problem.

The 1990s may see a greater acceptance of plastic parts in rugged conditions. The first automotive processors used ceramic parts to meet the environmental needs, particularly temperature and humidity. Today, plastic packaging technology has greatly improved so that plastic encapsulated parts are now used in these applications. With the emphasis of cost reduction within the military market, it is envisaged that plastic packages may be approved in preference to ceramic.

Processor technology

Microprocessors will use higher densities and faster speeds in two ways. The first will be to provide the fastest performance possible and the second is to provide the highest integration possible. In 1991, the general predictions for a state-of-the-art performance processor in 1995 one with over 15 million transistors and run at over 200 MHz. In 1994, the 200 MHz clock speeds are available with processor densities of about 3–4 million transistors. This may seem to be a failure in the predictions but it should remember that the technology is capable of such high integration but in many cases other factors have limited the use. Problems such as heat dissapation and timing delays involved in routing large number of signals across the day — many of today's designs are limited by transmission effects caused by bus impedances and capacitance — are preventing the use of the higher transistor counts.

At the other end of the scale, a 15 million transistor processor could integrate a complete desk top computer, or provide a system with 75 x MC68020 processors on it. It has even been suggested that a mixture of processors could be integrated to run different instruction sets! Integrated processors, like the M68300 family that has recently appeared, are only the beginning.

Memory technology

The Industry's memory technology plans indicate an exponential growth in memory size with a density doubling about every three years. DRAM chips tend to have a ten year design phase from initial process research through to peak demand. The recent peak demand for the 64 kbit and 256 kbit dynamic memory came from work that was started about ten years ago.

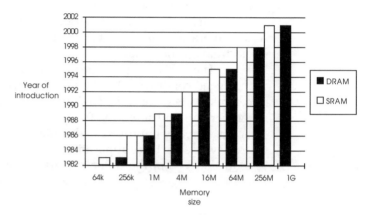

Memory technology predictions

DRAM memory life cycle falls into three stages. The first involves the development of the technology needed to fabricate the devices and to provide the wafer steppers and associated equipment needed to produce the designs on a large scale. The next stage introduces the parts and starts to ramp up production. This normally overlaps with the peak demand and subsequent decline of the preceding generations. This process is repeated throughout the generations. Static RAM follows a similar trend, although the memory densities are a factor of four smaller. Currently, 16 Mbit DRAMs and 4 Mbit SRAMs are generally available with larger sizes being sampled. The reduction in densities is explained by the increased number of transistors needed to build the cell — a DRAM cell uses one transistor while a SRAM cell uses four to six.

The effect of further reductions in the cost per bit is likely to cause more ripples within the computer and processor world. The most obvious is the introduction of solid state disks. Although this is happening today through the use of wafer scale integration, the cost reductions offered by having 32 Mbytes of RAM on a single chip will make this technology extremely accessible.

Other implications for system designers are equally interesting. Software designers are renowned for underestimating the amount of memory needed for software applications. With the growing trend of more user friendly interfaces and their insatiable demand for memory, tomorrow's personal computers will use many times the amount of memory compared to today´s. Newer technologies, like digital audio, will also benefit.

With 1 Gigabit of memory, it would be possible to store sufficient 16 bit words to play over 23 minutes of audio with compact disc quality! The data compression techniques provided by the new generations of processors could easily extend this playtime to hours. The days of revolving media, like records and tape cassettes, may be numbered. Again, the science fiction dream of a solid state audio cartridge is fast approaching reality.

Science fiction or not ?

For many, this glimpse into the future may appear to be too fantastic to believe. However, if the test of hindsight is applied to the last ten years, the developments seen in the eighties also appear as far fetched. Who would have predicted in 1980 that ten years later, most homes would have a computer complete with graphics and large amounts of memory and storage, that most cars would have more computing power in their ignition and control systems than a 1980 mini-computer could offer and that credit cards could act as complete data banks. Who would have predicted that the cost of an MC68000 16/32 bit processor would diminish to only 1% of its introductory cost! Some of the predictions are already coming true.

The Motorola-TRW announcement of their CPU-AX superchip, containing over 4 million transistors and calculating over 200 million floating point operations a second, reads like a science fiction novel. The die is 1.5" square and fits into a package 2.1" square and weighing 42 grams. The CMOS chip uses 0.5 micron technology and dissipates over 17 watts. The system consists of about 146 separate units — but only 64 need to function for it to perform correctly. Through a companion chip, the system can reconfigure itself and repair any faults that may occur during its fabrication or lifetime, providing this system minimum is maintained. The chip is the fruit of a ten-year-old VHSIC (very high scale integrated circuit) program sponsored by the American military.

This device seemed based on science fiction when it was announced in 1990 but several years on, the superscalar CISC processors like the MC68060 and Intel Pentium are approching the transister count and are not far from the performance levels. The signs are there to indicate that the future will be even more exciting than the past.

11   The changing design cycle

It is usual for a microprocessor system to be designed to meet commercial goals rather than for academic reasons and therefore part of the processor selection procedure has to take into account economic factors as well as engineering ones. The harsh realities of life dictate that the majority of processor designs must meet the required specification and be completed on time to stand any chance of commercial success. It is commercial success which pays everybody's wages and provides the resources for the next development.

The shortening design cycle

Perhaps the biggest change to the actual design process within recent years has been the shortening of both development cycle and product life. In 1984, the average microprocessor development took about 2 years from concept to preproduction. The product had a life of about 3 years before any major updating would be required. Today, the same development takes about 6 months, product life is about 18 months, with 2 or 3 mid-life upgrades built in! This contraction is caused by essentially one reason, hitting a market window, although there are several factors that are continuing to fuel these changes.

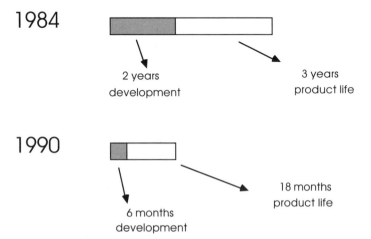

Design cycles

The importance of getting product to market at the right time is as great today as it has ever been. Competition within the world marketplace has increased product leapfrogging which, in turn, has caused product life reduction. The rapid introduction of new models to make competitive products obsolete, or appear to be, is an extremely potent commercial weapon. The fuel for these rapid developments is the pushdown effect, caused by exponentially expanding silicon technology. The pushdown effect refers to two characteristics of silicon technology. The first is that as new processors appear, the perceived performance of existing products goes down. This does not mean that an 8 MHz MC68000 is executing less instructions today than ten years ago, only that the expectations of a high performance processor in 1979 were met by a MC68000, while today an MC68040 is needed. Coupled with this are the unbelievable cost reductions that come with silicon technology. Comparing the cost of a digital watch in 1979 and today is an excellent example. For an example of this effect on processors, the MC68000 now costs only 1% of its introductory price and, in many cases, is now cheaper than 8 bit processors.

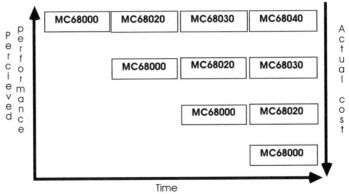

The pushdown effect

The follow-on effect is that products can be almost obsolete as soon as they appear and that they either have to be continually enhanced to remain in the high performance perception, or cost reduced to enter higher volume alternative markets. All this leads to the ever contracting design cycle.

These factors now have to be taken into account and they can pose considerable challenges to a design team. A combination of approaches may be needed. One is to take the new processors as soon as possible and continue to be on that leading edge of technology, thus being able to charge a premium for the product. A second may be to take a leading edge product but continue to cost reduce it so its growth comes from higher volume sales. The PC clone market is a good example of this tactic. A third could combine the two ideas so a particular aspect of the design can be provided on several different platforms, depending on price/performance etc. Whatever tactics are chosen, it is now almost mandatory to include provisions for these mid-life updates within the original design.

The double-edged sword of technology

For many projects, there will be a time when the double edged sword of technology[1] will have to be grasped. It is relevant to any design and usually refers to the time when a new technology or production method has to be developed or implemented. The learning curve involved is often underestimated, leading to the problems normally associated with new products. Many companies have policies of waiting for 12–24 months after a product is available before using it. As a result, the learning curve and its associated risk is greatly reduced, but so is any competitive advantage. Leading edge technology can be a tremendous help but it is equally good at cutting down to size. When successful, as SUN Microsystems found, the rewards are fabulous and quick. When not, the disaster can be equally as swift and final.

It is frequently important to gain early access to leading edge technology but the flipside of such access may be frequent design changes, the joy of being first (but if you have a problem, who do you ask for help?), delays, working without adequate documentation, development tools, etc. This type of environment is very exciting but it is really for the brave and stout-hearted!

For many, the way to solve the dilemma of getting access without the risk is through buying boards or complete systems rather than individual VLSI components. In the past, the argument of make versus buy has usually been based on economics — but within the framework of the changing design cycle, it has many other considerations.

Make versus buy

The economic argument is very simple. With low production runs, the most economic solution involves using product at its highest integration level. The diagram shows a graphical representation of this statement with reference to microprocessor systems. The crossover numbers on the bottom axis are approximations and depend ultimately on the individual situation. For runs of ten systems or less, it is generally better to buy in the complete system and simply add to it. For runs of 11–100 systems, it is more economic to buy in

---

[1] A phrase I first heard from Tim Coombs at Motorola.

boards and subassemblies and integrate the system. Higher than this, and it is better to design from scratch, buy in the silicon components and build everything.

The first statement that must be made, is that the crossover points are *extremely* woolly. They involve balancing the costs of increased work against the combined component savings and this can frequently change both from situation to situation and with time.

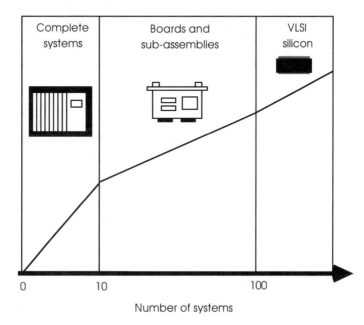

Number of systems

Make versus buy

This argument does not take into account other factors which change on moving from one design mode to another.

Design flexibility

Using systems for low production runs has some drawbacks. The design will have compromised the functional need for using an off-the-shelf product. This has several potential follow ons. The first is that the system may need modifications which could cost more than the savings. The second concerns system architecture. An off-the-shelf system may have all the functionality needed (i.e. processor, memory, peripherals, etc.) but the overall system architecture may not be exactly suitable for the application. A third potential problem concerns future enhancements or cost reductions. Using a proprietary end-user system may restrict such upgrades and thus cause difficulties later. It may not possible to obtain the system with unused functions removed. At this point, the solution may be to move to a board solution.

Building a system using an industry standard bus, like VMEbus or Multibus II, has become an extremely common solution and has led to many hundreds of suppliers providing an extremely wide range of products. This has allowed many companies to buy a core system, develop any specialised boards and integrate the complete system. By using such buses, upgrades are relatively simple and only involve board swaps. The resultant design is more closely matched to needs, although some trade-offs at a lower level may still have to be made. The downside is increased costs and a potentially longer design cycle — but for many, this is a good way of building or developing a product.

If the design has special architectural or build needs, such as a special multiprocessing architecture or military approval, the compromises needed to be able to take advantage of boards or systems may not be acceptable. The only solution is to build the system from scratch. The costs of doing so should not be underestimated and many prototype systems have not seen the light of day because of increased workload. Where the risk is seen as high, it may be better to compromise the design slightly and use bought in subassemblies to enable a prototype to be built. In some cases, this approach has enabled companies to meet early commitments when the full custom system has suffered from delays.

Manufacturing costs

With the move from a system through boards to silicon, manufacturing costs increase dramatically. These costs include the increased development times, inventory costs, capital equipment, etc. and should not be underestimated! Unfortunately, there is often a passionate plea from engineers to design a whole system, rather than concentrating on building the special functions and buying in the more standard parts, such as memory and processor boards.

There are cases when manufacturing from silicon is viable for small production runs. The most obvious is when the required function is simply not available off the shelf and has to be built. However, do not be lulled into adopting the approach of 'building a board' and justifying its cost by selling it as an off-the-shelf product. This rarely works. With a VMEbus based product, there are over 300 vendors supplying product and this competitive market is offered nearly every function needed. To successfully sell a product into this type of environment requires additional software support, service support, etc., which increases the costs or reduces profit margins. In addition, the marketing required is totally different and is like selling a commodity rather than high technology.

Military approval is another case where the goal of high reliability overrides the normal economic arguments. Even here, this is not as clear cut as it used to be. Many military systems now use commercial product within a sealed environmental chamber so that costs can be reduced but the operational specification is not. With the recent advent of fixed price contracts, this approach is becoming more common.

The availability of manufacturing facilities and the need to keep them fully loaded is another exception. As system complexity increases, designs are frequently being licensed allowing boards to be used for prototyping but basing the production runs on component costs. The licence deal removes the design and development phases.

Quality

Quality has become paramount in recent years and, again, achievable levels can depend on how the product has been engineered. With systems, reliability is dictated by the product quality provided by the system supplier. If the product is bad, it does not matter how good the added value is, the overall system will be deemed as unreliable and unacceptable. The same comment applies to board products however, system integration and testing is under control and this can allow better screening to improve quality, albeit at a cost. At component level, acceptance testing is expensive, requiring sophisticated testers and test program development. Many users are reducing their supplier base to those whose quality is such that the product can be shipped straight to the production line without interim testing.

Upgrading

The opening paragraphs of this chapter described how today's product life was often about 18 months, with 2 or 3 mid-life upgrades to ensure the length. It is important to design these upgrades into the system specification from day one.

With systems, the upgrade policy is simple to implement — simply go and buy the latest 'go faster' version and transfer the system software and special hardware over. This is a relatively painless process but it does depend on two factors — the compatibility between the systems and, secondly, any hardware dependency within the application software. The most common culprits are software loops, which no longer work due to increased processor performance, and the failure to use defined procedures to access physical I/O. It is advisable when designing a system to add an upgrade check to the approval process. It is a good rule to assume nothing, or alternatively, document all assumptions and their implications.

The same scenario is true for board designs. Different speed processor boards can often be interchanged to offer the appropriate performance as needed. However, boards offer greater flexibility and therefore there is an added risk of the production configuration being the one which works on paper but does not work in reality. These problems often occur when functions are added at the same time as performance is upgraded. An onboard Ethernet controller may not be compatible with its separate predecessor or an onboard disk controller may not be able to use existing disk drives that were compatible with the previous discrete controller. These implications can often incur additional costs within the rest of the system which cancel out the advantages of changing in the first place. To ensure these problems do not appear in the field, sufficient resources must be allocated to test any new configurations. Again the 'how will this affect an upgrade' test is invaluable.

For silicon designs, the cost of making an unplanned change involving a modification to the printed circuit board or test programs can be very large indeed and should therefore be avoided. Upgrades for discrete designs are fairly limited in scope, compared with board and systems, and invariably have to be catered for at the design inception. Typical techniques include designing the system to support several different processor speeds, memory sizes and speeds and the provision of an expansion port to allow more functionality. Upgrades then become a simple matter of adding different or extra components.

Product differentiation

The question of product differentiation is quite esoteric and primarily concerned with how different a product is and how it stands within a marketplace. Within the electronics industry, there is a large amount of badge engineering where systems are purchased from a supplier by many different companies, who then change the front badge and re-sell the system. In this case, product differentiation is not the product itself — it is the same from each supplier — but the added value, in terms of service backup, application software, networking or just the brand name! When a prospective client can see that the system (well, the hardware anyway) is the XYZ Desktop available from any discount store, it is difficult to prevent incorrect comparisons being made. The key is to justify the extra cost by the added value, so objections like 'why should I pay five times for the system for a bit of software' can be countered.

With boards, there is a little more that can be done to differentiate it from other versions of it. Making drastic changes to the case etc. is expensive and more indicative of adopting the board or sub-assembly approach. There is more scope to actually changing the physical appearance and differentiation. Corporate colours and footprints can be adopted to emphasise its presence within an integrated range of products. Although the product may be functionally compatible with the XYZ Desktop, it is harder to make the comparisons and thus raise objections.

With a discrete design, the choices concerning appearance, performance and functionality are not restrained. The product differentiation is at its greatest.

Service and backup

This is often a neglected part of the equation. Despite the best efforts of designers, production and quality staff, products do occasionally fail or require regular maintenance. This needs to be budgeted for. For systems, the original manufacturer´s service organisation can be used to provide the backbone of the service, with only hardware specials and software to be catered for. This can save a lot of money and be a major reason for adopting a system solution for even high run rates. The types of services offered are usually based on a 24 hour on-site repair scheme.

With boards, more responsibility must be taken by the end manufacturer. Board manufacturers often offer a 24 hour replacement by post or courier service, which allows the end manufacturer to maintain a small inventory of spares and provide a higher level of service. This obviously increases costs.

With a discrete system, the design and implementation is in the hands of the end manufacturer and, therefore, the entire service costs are his responsibility.

Production technology

The production tools needed range from screwdrivers for a system implementation, to a full production line with PCB design facilities etc. These costs can be quite prohibitive and, in reality, much use is made of sub-contractors for PCB manufacture, board stuffing and testing. However, the implicit costs of using certain types of packaging or achieving certain component densities are still there, irrelevant of in house or outside manufacturing.

The problem concerns the ever increasing trend of more pins within smaller areas running at faster speeds. The first 8 bit micros had a state-of-the-art pin count of about 40 pins in a dual in-line package based on a 600 mil pitch. The M68000 went up to 64 pins in a dual in-line package which occupied about 3 square inches. To many, it was affectionately known as a caterpillar! The same device is now available in a surface mount FN package which is less than 0.75 square inches in area. The MC68020 came out as a large pin grid array package with 114 pins and, for many PCB designers, was a nightmare come true. It is now accepted as a standard package, with the nightmare being a surface mount package with over 200 leads on a 25 thousandths of an inch pitch! What is important to remember is that the layout tools have improved, so that while the new packages can present a challenge, they are still within the current technology. The decision has then to be made whether this challenge is accepted or deferred until later, when implementation is easier.

The same comments are true for PCB production. A four-layer board with two power and two signal planes was once state of the art. Fourteen-layer boards are now commonplace, with six power planes and eight signal planes. Track widths have reduced so that two tracks can be laid between adjacent pins. This type of technology is needed to make the large numbers of connections present in a modern high density board. Coupled with this need for more connections, is the increase in processor clocks, which forces more care to be taken to prevent transmission line effects from distorting previously clean edges.

The pushdown effect previously described for processors also hold true for production technology.

Simulation versus emulation

The methods used to develop and debug systems within the design cycle have also had to change to meet the new requirements.

The in-circuit emulator has been the usual method used for debugging microprocessor systems. It gives basic facilities for software testing, hardware debugging and system integration by providing a window into the system via the processor. Its ability to simulate a simple target (i.e. a processor with memory) has allowed software to be developed and tested in parallel with the hardware. The target hardware was then debugged using the emulator to ensure that the basic functions worked before the task of total system integration was attempted.

However, emulators are not the perfect replacements for microprocessor silicon. The logic needed to intercept processor signals, substitute functions and drive other signals causes slight timing differences. These differences are further aggravated by the delays caused in moving signals up and down the cable connecting the probe between the target and the emulator. With processor pin counts fast approaching the 200 pin barrier and beyond (resulting in large probes), the target board must often be laid out to accommodate the emulator rather than the processor! Surface mount packages and the ever decreasing pin width are further obstacles to the use of emulators.

As systems have become more complex, the associated problems have required logic analyser functions to provide a wider picture than that of the processor environment. To solve these problems, data must be provided by the processor and other functions to give an overall view. This data often needs further qualification to isolate the critical events from the background. With the promise of faster and more powerful processors, how are such systems going to be developed?

The answer is to provide better tools earlier in the design cycle. The emulator has traditionally been used to debug hardware, identify any prototyping problems and eventually provide a stable hardware target. With the high cost of complex multilayer circuit boards, the cost of any modification is high. This draws a parallel with ASIC development — in this environment, mistakes cannot be simply solved by the addition of blue wires — the designer has to get it right. With ASIC design, the key to achieving this is the right simulation tools.

Motorola realised that the new generation of processors, like the M88000 family, was going to offer single cycle performances with fast clocks (20–33 MHz) where emulation would be virtually impossible to use (this problem was even worse with the introduction of the PowerPC with 60, 66 and 80 MHz parts). Designers are already using the fastest logic and keeping PCB track lengths to a minimum to reduce transmission line effects and delays. With their inherent timing differences, emulators are not suitable. The problems with noise and crosstalk effectively prevent wirewrapping, and it has become apparent that many customers were going straight to PCB for prototyping. The multilayer boards involved to achieve the density and noise immunity are expensive and it was therefore essential to get the design right first time. The solution was simple — take the ASIC example and develop simulation tools to allow both the hardware and software to be simulated and tested early in the design.

The first tool was a software simulator which could allow cross-compiled code to be tested and debugged on a basic system. This allowed software development in parallel with the target design. The M88000 simulator provides symbolic level debugging, single stepping, register and memory display and modification, breakpoints, etc. It has been integrated with the Greenhills compilers by Oasys and is available from Real Time Products. To help speed this development, Motorola recently placed a debug monitor into the public domain for both the M68000 and M88000 processor families. The debugger, FBUG, is written in 'C' and can easily be configured for any system.

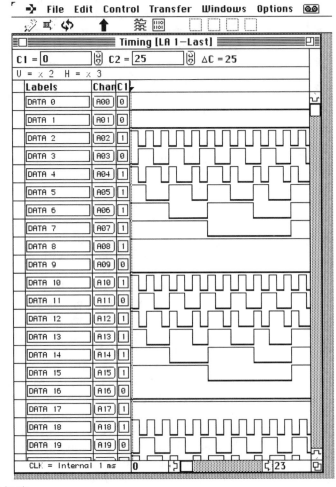

*A Gould CLAS4000 logic
analyser timing display*

File Edit Control Transfer Windows Options

State [LA 1—Last]

C1=0 C2=25 C2-C1= 0025 samples Previous Next

C	Line	Address	Data	Status		C	Line	Address	Data	Status
C1	0000	F5F4	F7F6	F9F8			0037	3938	3B3A	3D3C
	0001	E9E8	EBEA	EDEC			0038	2D2C	2F2E	3130
	0002	DDDC	DFDE	E1E0			0039	2120	2322	2524
	0003	D1D0	D3D2	D5D4			0040	1514	1716	1918
	0004	C5C4	C7C6	C9C8			0041	0908	0B0A	0D0C
	0005	B9B8	BBBA	BDBC			0042	FDFC	FFFE	0100
	0006	ADAC	AFAE	B1B0			0043	F1F0	F3F2	F5F4
	0007	A1A0	A3A2	A5A4			0044	E5E4	E7E6	E9E8
	0008	9594	9796	9998			0045	D9D8	DBDA	DDDC
	0009	8988	8B8A	8D8C			0046	CDCC	CFCE	D1D0
	0010	7D7C	7F7E	8180			0047	C1C0	C3C2	C5C4
	0011	7170	7372	7574			0048	B5B4	B7B6	B9B8
	0012	6564	6766	6968			0049	A9A8	ABAA	ADAC
	0013	5958	5B5A	5D5C			0050	9D9C	9F9E	A1A0
	0014	4D4C	4F4E	5150			0051	9190	9392	9594
	0015	4140	4342	4544			0052	8584	8786	8988
	0016	3534	3736	3938			0053	7978	7B7A	7D7C
	0017	2928	2B2A	2D2C			0054	6D6C	6F6E	7170
	0018	1D1C	1F1E	2120			0055	6160	6362	6564
	0019	1110	1312	1514			0056	5554	5756	5958
	0020	0504	0706	0908			0057	4948	4B4A	4D4C
	0021	F9F8	FBFA	FDFC			0058	3D3C	3F3E	4140
	0022	EDEC	EFEE	F1F0			0059	3130	3332	3534
	0023	E1E0	E3E2	E5E4			0060	2524	2726	2928
	0024	D5D4	D7D6	D9D8			0061	1918	1B1A	1D1C
C2	0025	C9C8	CBCA	CDCC			0062	0D0C	0F0E	1110
	0026	BDBC	BFBE	C1C0			0063	0100	0302	0504
	0027	B1B0	B3B2	B5B4			0064	F5F4	F7F6	F9F8
	0028	A5A4	A7A6	A9A8			0065	E9E8	EBEA	EDEC
	0029	9998	9B9A	9D9C			0066	DDDC	DFDE	E1E0
	0030	8D8C	8F8E	9190			0067	D1D0	D3D2	D5D4
	0031	8180	8382	8584			0068	C5C4	C7C6	C9C8
	0032	7574	7776	7978			0069	B9B8	BBBA	BDBC
	0033	6968	6B6A	6D6C			0070	ADAC	AFAE	B1B0
	0034	5D5C	5F5E	6160			0071	A1A0	A3A2	A5A4
	0035	5150	5352	5554			0072	9594	9796	9998
	0036	4544	4746	4948			0073	8988	8B8A	8D8C

A Gould CLAS4000 logic analyser hexa-decimal decoded display

A second set of simulation tools was developed in conjunction with Logic Automation. This provides hardware models for Motorola´s processors and peripherals and allows complete target systems to be designed and simulated. System timings can be checked to indentify any errors. Smartmodel simulations allow software to be executed on the system and can be used to integrate software with the hardware model without actually having the real physical hardware available. There are several advantages to this: test software can be debugged prior to integrtaion with the hardware and thus help resolve the eternal question of whether the fault is due to faulty software or hardware. New design ideas can be tested on a 'what if' basis. The emphasis is on solving problems when modifications are inexpensive to carry out, i.e. early in the design cycle.

Once simulated, the basic target board can be built and the full software tested. Compared with previous generations, today´s systems are more complex and often feature multiprocessors with cache memory systems etc. Many of the problems that now occur are subsystem interactions, where either communication or synchronisation has failed. Identifying such problems with only an emulator is like looking at a room via a keyhole. It provides a view into the system but one cannot see everything or be sure that something isn´t lurking out of sight. In an effort to widen this view, many emulators now provide basic logic analysis support or, in the case of the MC68302 ADS, special decoded processor signals for logic analysers. Simple analysers have their disadvantages—while they can unobtrusively collect vast quantities of data, they often cannot display it in a ideal format, such as op codes.

The solution is to provide the new generation of analysers with a disassembly capability and sophisticated event qualification, such as those from Gould, Hewlett Packard and Tektronix. These analysers provide very large numbers of channels, essential for a processor with 4 x 32 bit wide ports, which can disassemble data streams either side of the MC88200 cache memory management unit. In this way, both logical and physical bus signals can be combined with other data signals to provide the database for further event qualification. This can be so powerful that it is easy to isolate even erratic random failures and trace the before and after status. The general purpose nature of this instrumentation is another benefit—it can support a wide range of processors and will analyse data from any system. To switch from M68000 to PowerPC disassembly is often a simple software module change.

These techniques are not the abrupt end of the emulator. For slower speeds and less complex designs, it can still provide an efficient debugging environment. Both the recently announced Motorola MC68332 and MC68302 have in-circuit emulation tools and, as emulation technology improves, undoubtedly systems will appear for faster processors. However, technology is increasing exponentially. Observers are still predicting that 200 MHz processors with 512+ connections and with the chip directly bonded to a substrate will be here before the turn of the century. Only time will tell whether emulation or simulation techniques will have kept pace with the current technology.

High level language simulation

If software is written in a high level language, it is possible to test large parts of it without the need for the VMEbus hardware at all. Software that does not need or use I/O or other system dependent facilities can be run and tested on other machines, such as a PC or a engineering workstation. The advantage of this is that it allows a parallel development of the hardware and software and added confidence when the two parts are integrated that it will work.

Using this technique, it is possible to simulate I/O using the keyboard as input or another task passing input data to the rest of the modules. Another technique is to use a data table which contains data sequences that are used to test the software.

This method is not without its restrictions. The most common mistake with this method is the use of non-standard libraries which are not supported by the target system compiler or environment. If these libraries are used as part of the code that will be transferred, as opposed to providing a user interface or debugging facility, then the modifications needed to port the code will devalue the benefit of the simulation.

The ideal is when the simulation system is using the same library interface as the target. This can be achieved by using the target system or operating system as the simulation system or using the same set of system calls. Many operating systems support or provide a UNIX-compatible library which allows UNIX software to be ported using a simple re-compilation. As a result, UNIX systems are often employed in this simulation role. This is an advantage which the POSIX compliant operating system Lynx offers.

This simulation allows logical testing of the software but rarely offers quantitative information unless the simulation environment is very close to that of the target, in terms of hardware and software environments.

Low level simulation

Using another system to simulate parts of the code is all well and good, but what about low level code such as initialisation routines? There are simulation tools available for these routines as well. CPU simulators can simulate a processor, memory system and, in some cases, some peripherals and allow low level assembler code and small HLL programs to be tested without the need for the actual hardware. These tools tend to fall into two categories: the first simulate the programming model and memory system and offer simple debugging tools similar to those found with an on board debugger. These are inevitably slow, when compared to the real thing, and do not provide timing information or permit different memory configurations to be tested. However, they are very cheap and easy to use and can provide a low cost test bed for individuals within a large software team. There are even shareware MC68000 simulators available such as the one from the University of North Carolina which simulates a MC68000 processor.

```
<D0> =00000000 <D4> =00000000 <A0> =00000000 <A4> =00000000
<D1> =00000000 <D5> =0000abcd <A1> =00000000 <A5> =00000000
<D2> =00000000 <D6> =00000000 <A2> =00000000 <A6> =00000000
<D3> =00000000 <D7> =00000000 <A3> =00000000 <A7> =00000000
trace: on      sstep: on      cycles:    416     <A7'>= 00000f00
          cn tr st rc         T S  INT   XNZVC  <PC> = 00000090
   port1  00 00 82 00  SR = 1010101111011111
--------------------------------------------------------
executing a ANDI          instruction at location   58
executing a ANDI          instruction at location   5e
executing a ANDI          instruction at location   62
executing a ANDI_TO_CCR   instruction at location   68
executing a ANDI_TO_SR    instruction at location   6c
executing a OR            instruction at location   70
executing a OR            instruction at location   72
executing a OR            instruction at location   76
executing a ORI           instruction at location   78
executing a ORI           instruction at location   7e
executing a ORI           instruction at location   82
executing a ORI_TO_CCR    instruction at location   88
executing a ORI_TO_SR     instruction at location   8c
TRACE exception occurred at location    8c.
Execution halted
```

*Example display from
the University of North
Carolina 68k simulator*

The second category extends the simulation to provide timing information based on the number of clock cycles. Some simulators can even provide information on cache performance, memory usage and so on, which is useful data for making hardware decisions. Different performance memory systems can be exercised using the simulator to provide performance data. This type of information is virtually impossible to obtain without using such tools. These more powerful simulators often require very powerful hosts with large amounts of memory.

Simulation tools are becoming more and more important in providing early experience of and data about a system before the hardware is available. They can be a little impractical due to their performance limitations — one second of a processing with a 25 MHz RISC processor can take 2 hours of simulation time! — but as workstation performance improves, so their use will increase.

Onboard debugger

The onboard debugger provides a very low level method of debugging software. Supplied as a set of EPROMs which are plugged into the board and using a serial port to communicate with a terminal, it performs two functions. The first is to provide some basic software for the board to run, which will normally initialise the hardware and allow it come up into a known state. The second is to supply basic debugging facilities and, in some cases, allow simple access to the board's peripherals. Often included in these facilities is the ability to download code using a serial port or from a floppy disk.

When the board is powered up, the processor fetches its reset vector from the table stored in EPROM and then starts to initialise the board. The vector table is normally transferred from EPROM into a RAM area to allow it to be modified, if needed. This can be done through hardware, where the EPROM memory address is temporarily altered to be at the correct location for power on, but is moved elsewhere after the vector table has been copied. Typically, a counter is used to determine a pre-set number of memory accesses, after which it is assumed that the table has been transferred by the debugger and the EPROM address can safely be changed.

```
>TR

PC=000404 SR=2000 SS=00A00000 US=00000000          X=0
A0=00000000 A1=000004AA A2=00000000 A3=00000000 N=0
A4=00000000 A5=00000000 A6=00000000 A7=00A00000 Z=0
D0=00000001 D1=00000013 D2=00000000 D3=00000000 V=0
D4=00000000 D5=00000000 D6=00000000 D7=00000000 C=0
---------->LEA        $000004AA,A1

>TR

PC=00040A SR=2000 SS=00A00000 US=00000000          X=0
A0=00000000 A1=000004AA A2=00000000 A3=00000000 N=0
A4=00000000 A5=00000000 A6=00000000 A7=00A00000 Z=0
D0=00000001 D1=00000013 D2=00000000 D3=00000000 V=0
D4=00000000 D5=00000000 D6=00000000 D7=00000000 C=0
---------->MOVEQ     #19,D1

>
```

Example display from an onboard M68000 debugger

The second method, which relies on processor support, allows the vector table to be moved elsewhere in the memory map. With the later M68000 processors, this can also be done by changing the vector base register which is part of the supervisor programming model.

The debugger usually operates at a very low level and allows basic memory and processor register display and change, setting RAM based breakpoints and so on. This is normally performed using hexadecimal notation, although some debuggers can provide a simple disassembler function. To get the best out of these systems, it is important that a symbol table is generated when compiling or linking software, which will provide a cross reference between labels and symbol names and their physical address in memory. In addition, an assembler source listing which shows the assembler code generated for each line of C or other high level language code is invaluable. Without this information it can be very difficult to use the debugger easily. Having said that, it is quite frustrating having to look up references in very large tables and this highlights one of the restrictions with this type of debugger.

While considered very low level and somewhat limited in their use, onboard debuggers are extremely useful in giving confidence that the board is working correctly and working on an embedded system where an emulator may be impractical. However, this ability to access only at a low level can also place severe limitations on what can be debugged.

The first problem concerns the initialisation routines and in particular the processor's vector table. Breakpoints use either a special breakpoint instruction or an illegal instruction to generate a processor exception when the instruction is executed. Program control is then transferred to the debugger which displays the breakpoint and associated information. Similarly, the debugger may use other vectors to drive the serial port that is connected to the terminal.

This vector table may be overwritten by the initialisation routines of the operating system which can replace them with its own set of vectors. The breakpoint can still be set but when it is reached, the operating system will see it instead of the debugger and not pass control back to it. The system will normally crash because it is not expecting to see a breakpoint or an illegal instruction!

To get around this problem, the operating system may need to be either patched so that its initialisation routine writes the debugger vector into the appropriate location or this must be done using the debugger itself. The operating system is single stepped through its initialisation routine and the instruction that overwrites the vector simply skipped over, thus preserving the debugger's vector.

A second issue is that of memory management where there can be a problem with the address translation. Breakpoints will still work but the addresses returned by the debugger will be physical, while those generated by the symbol table will normally be logical. As a result, it can be very difficult to reconcile the physical address information with the logical information.

The onboard debugger provides a simple but sometimes essential way of debugging VMEbus software. For small amounts of code, it is quite capable of providing a method of debugging which is effective, albeit not as efficient as a full blown symbolic level debugger — or as complex or expensive. It is often the only way of finding out about a system which has hung or crashed.

Task level debugging

In many cases, the use of a low level debugger is not very efficient compared with the type of control that may be needed. A low level debugger is fine for setting a breakpoint at the start of a routine but it cannot set them for particular task functions and operations. It is possible to set a breakpoint at the start of the routine that sends a message, but if only a particular message is required, the low level approach will need manual inspection of all messages to isolate the one that is needed — an often daunting and impractical approach!

To solve this problem, most operating systems provide a task level debugger which works at the operating system level. Breakpoints can be set on system circumstances, such as events, messages, interrupt routines and so on, as well as the more normal memory address. In addition, the ability to filter messages and events is often included. Data on the current executing tasks is provided, such as memory usage, current status and a snapshot of the registers.

Symbolic debug

The ability to use high level language instructions, functions and variables instead of the more normal addresses and their contents is known as symbolic debugging. Instead of using an assembler listing to determine the address of the first instruction of a C function and using this to set a breakpoint, the symbolic debugger allows the breakpoint to be set by quoting a line reference or the function name. This interaction is far more efficient than working at the assembler level, although it does not necessarily mean losing the ability to go down to this level if needed.

The reason for this is often due in the way that symbolic debuggers work. In simple terms, they are intelligent front ends for assembler level debuggers, where software performs the automatic lookup and conversion between high level language structures and their respective assembler level addresses and contents. The key to this is the creation of a symbol table which provides the cross-referencing information that is needed. This can either be included within the binary file format used for object and absolute files or, in some cases, stored as a separate file. The important thing to remember is that symbol tables are often not automatically created and, without them, symbolic debug is not possible.

When the file or files are loaded or activated by the debugger, it searches for the symbolic information which is used to display more meaningful information as shown in the various listings. The symbolic information means that breakpoints can be set on language statements as well as individual assembler addresses. Similarly, the code can be traced or stepped through line by line or instruction by instruction.

```
12              int     prime,count,iter;
13
14              for (iter = 1;iter<=MAX_ITER;iter++)
15                  {
16                  count = 0;
17                  for(i = 0; i<MAX_PRIME; i++)
18                      flags[i] = 1;
19                  for(i = 0; i<MAX_PRIME; i++)
20                      if(flags[i])
21                      {
22                        prime = i + i + 3;
23                        k = i + prime;
24                        while (k <
MAX_PRIME)
25                        {
26                          flags[k] = 0;
27                          k += prime;
28                          }
29                        count++;
```

Source code listing
with line references

```
›  000100AA 7C01            MOVEQ    #$1,D6
›  000100AC 7800            MOVEQ    #$0,D4
>  000100AE 7400            MOVEQ    #$0,D2
›  000100B0 207C 0001 2148  MOVEA.L  #$12148,A0
›  000100B6 11BC 0001 2000  MOVE.B   #$1,($0,A0,D2.W)
›  000100BC 5282            ADDQ.L   #$1,D2
›  000100BE 7011            MOVEQ    #$11,D0
›  000100C0 B082            CMP.L    D2,D0
›  000100C2 6EEC            BGT.B    $100B0
›  000100C4 7400            MOVEQ    #$0,D2
›  000100C6 207C 0001 2148  MOVEA.L  #$12148,A0
›  000100CC 4A30 2000       TST.B    ($0,A0,D2.W)
›  000100D0 6732            BEQ.B    $10104
```

```
› 000100D2 2A02              MOVE.L   D2,D5
› 000100D4 DA82              ADD.L    D2,D5
› 000100D6 5685              ADDQ.L   #$3,D5
```

Assembler listing

This has several repercussions. The first is the number
of symbolic terms and the storage they require. Large tables
can dramatically increase file size and this can pose constraints on linker operation when building an application or
a new version of an operating system. If the linker has insufficient space to store the symbol tables while they are being
corrected — they are often held in RAM for faster searching
and update — the linker may crash with a symbol table
overflow error. The solution is to strip out the symbol tables
from some of the modules by recompiling them with symbolic
debugging disabled or by allocating more storage space to the
linker.

```
›>>    12       int     prime,count,iter;
›>>    13
›—     14  =>       for (iter = 1;<=iter<=MAX_ITER;iter++)
› 000100AA 7C01                  MOVEQ    #$1,D6
›>>    15          {
›>>    16          count = 0;
› 000100AC 7800                  MOVEQ    #$0,D4
›—     17  =>          for(i = 0;<=  i<MAX_PRIME; i++)› >
000100AE 7400                   MOVEQ    #$0,D2
›>>    18          flags[i] = 1;
› 000100B0 207C 0001 2148     MOVEA.L  #$12148,A0
{flags}
› 000100B6 11BC 0001 2000      MOVE.B   #$1,($0,A0,D2.W)
›—     17              for(i = 0; i<MAX_PRIME; => i++)<=
› 000100BC 5282                  ADDQ.L   #$1,D2
›—     17              for(i = 0; => i<MAX_PRIME;<=i++)
› 000100BE 7011                  MOVEQ    #$11,D0
› 000100C0 B082                  CMP.L    D2,D0
› 000100C2 6EEC                  BGT.B    $100B0
```

*Assembler listing with
symbolic information*

The problems may not stop there. If the module is then
embedded into a target and symbolic debugging is required,
the appropriate symbol tables must be included in the build
and this takes up memory space. It is not uncommon for the
symbol tables to take up more space than the spare system
memory and prevent the system or task from being built or
running correctly. The solution is to add more memory or
strip out the symbol tables from some of the modules.

It is normal practice to remove all the symbol table information from the final build to save space. If this is done, it will also remove the ability to debug using the symbol information. It is a good idea to at least have a hard copy of the symbol table to help should any debugging be needed.

Emulation

Even using the described techniques, it cannot be stated that there will never be a need for additional help. There will be times when instrumentation, such as emulation and logic analysis, is necessary to resolve problems within a design quickly. Timing and intermittent problems cannot be easily solved without access to further information about the processor and other system signals. Even so, the recognition of a potential problem source, such as a specific software module or hardware, allows more productive use and a speedier resolution. The adoption of a methodical design approach and the use of ready built VMEbus boards as the final system, at best remove the need for emulation and, at worst, reduce the amount of time debugging the system.

12 The next generations

MC68060 — superscalar CISC

The MC68060 is a superscalar M68000 compatible processor capable of executing two instructions per clock. Its internal architecture as shown in the diagram is very similar to that of a RISC design with Harvard data and instruction caches feeding multiple execution units. The execution units are heavily pipelined and feature branch folding and separate write back stages — similar to that used in the PowerPC architecture.

The key to its operation is in the instruction decoding. As previously discussed, the variable size of the instruction and its complicated decoding cause a lot of difficulties for superscalar design. To get around this restriction, the M68000 instructions are taken in from the instruction cache and are internally converted into a fixed-size instruction format. This internal representation is very similar that used by a superscalar RISC processor.

This instruction conversion is combined with a branch folding mechanism and branch address cache to resolve and remove branch instructions before they enter the primary and secondary instruction units.

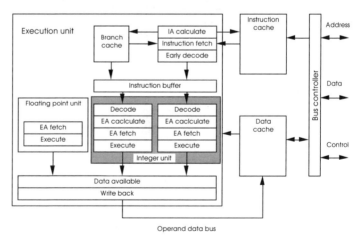

MC68060 internal block diagram

The design can execute either two integer instructions per clock or an integer with a floating point instruction or branch instruction.

Instruction fetch unit

This consists of several stages and is where the instruction format is converted and any branch instructions resolved.

- Instruction address calculation
  This is where the virtual address of the instruction is calculated to enable it to be fetched form memory.
- Instruction fetch
  This is where the instruction is fetched from memory.
- Early decode
  This is where the instruction is partially decoded and converted into its fixed size internal representation.
- Instruction buffer
  This contains the converted instructions, awaiting dispatch to the execution units.

Branch unit

This unit resolves the branch using an internal branch prediction cache and through branch folding. For each intstruction address that is generated in the pipelines, it is checked against the contents of the branch cache. If there is no hit, the pipeline continues to process instructions sequentially as normal. If there is a hit, then the instruction stream is discarded and replaced with with a new one using the address stored in the branch cache. With the instruction buffering, the instruction stream that is presented to the execution unit has the branches removed and the instruction stream appears to be sequential.

The MPC604

The MPC604 is the third PowerPC processor to come from the joint IBM–Motorola–Apple design team based in Austin, Texas. It continues the superscalar nature of the PowerPC architecture by increasing the number of instructions that it can execute per clock to four. To do this it has six internal execution units: two single-cycle integer units, a multiple-cycle integer unit, a floating point unit, a branch processing unit and a load store unit. Although four instructions are issued or retired on every clock edge, it is possible for all six units to be working on instructions and thus process as many as six instructions at any one time.

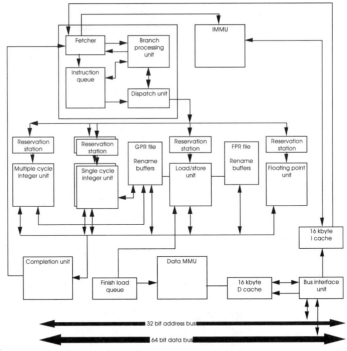

The MPC604 block
diagram

The future for CISC

The ability for CISC processor architectures to keep pace with the performance improvements offered by RISC is becoming less and less important. The indications are that the days of CISC processors driving the leading edge PCs and workstations are coming quickly to an end. RISC architectures are at last delivering their promise of better performance with less power dissapation and using less silicon. The reason for this change is not solely due to the advantages offered by the new architectures and their hardware design but also through their ability to support legacy software.

With software becoming the dominant cost in any computer system, the benefit of a bigger, faster alternative is lost if new software needs to be bought immediately to take advantage of the new system. The ability through emulation to run this legacy software is now a key part of the scenario needed for success in the PC and workstation market. One of the problems faced with this type of technology is gaining sufficient processor performance to compensate for the software emulation overhead. With the advantages offered by

RISC, it is now possible to combine the two technologies and thus provide a way of migrating legacy software to other platforms and processor architectures.

What does this mean for CISC processors like the M68000 and Intel 8086 architectures? Both Motorola and Intel have indicated that for the best performance, RISC offers the best route foreward and that emulation technology can be used to provide a migration path. For Motorola, that path has already been defined through the collaboration with IBM and Apple to produce PowerPC. This alliance, announced in 1991 has now resulted in the first three members of the processor family: the MPC601, 603 and 604. Intel have just announced an alliance with Hewlett Packard to provide a RISC-based successor to the 80x86 architecture. So certainly the die is cast to move to RISC.

The real future for the M68000 family is in the integrated processor for use within embedded systems such as car electronics, office peripherals, communication devices and so on. This is not something that has suddenly appeared: integrated processors like the MC68302 have been available for several years. What has happened is the generation of a complete family. What has happened to the 8 bit microcontroller where one or two initial designs have led to hundreds of derivatives has now happened to the M68000 family.

An alternative direction — system integration

For many applications, the promise of higher performance does not, in itself, satisfy the design needs. A design may need low chip counts and small packaging instead. Fortunately, the silicon technology which has increased clock speeds and device density to provide today´s high performance processors can equally be used to integrate a distributed processor system on to one chip with its reduced chip count and costs.

The M68300 family

This family of integrated M68000 processors bridges the gap between 8 bit microcontrollers and the 16/32 bit M68000 families. The idea was simple — take an M68000 family processor core, add all the interface logic needed for chip select, some onchip memory and some intelligent peripherals controlled by another dedicated processor. This type of configuration is almost identical to that of an MC68HC11 MCU, except that processing power is greatly increased.

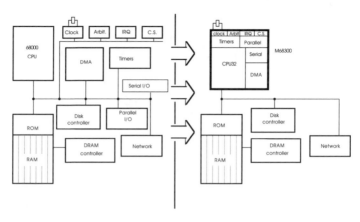

The basic integration principle behind the MC68300 family

The first part, the MC68332, was developed for General Motors to satisfy its automotive requirements. The invasion of the microprocessor into this industry has been enormous in recent years. Microprocessors now control ignition and fuel injection systems, anti-lock braking, gearbox control, dashboard instrumentation and even folding roofs in some up-market convertibles.

These functions are now being performed by essential 8 bit and derivative architectures and an immediate question comes to mind — 'why move to a 16/32 bit architecture ?'

The answer is simple. Microprocessors allow finer control of the systems and can compensate for mechanical wear and tear within a mechanical component's life. Both these attributes are essential for an engine to maintain peak performance and, more importantly, emission control. They reduce the amount of servicing needed which, in turn, improves reliability and decreases costs. However, if they fail, there is little a motorist can do to rectify a fault. A modern car has no distributor, optical timing sensors, integrated high tension supply for the spark plugs and so on. Even the electronics are encapsulated! The continuous need to improve environmental aspects without increasing costs demands more and more processing power.

The diagram shows a simplified engine management system and identifies many of the tasks performed by the processor within the system. The work load is even greater for 6, 8 or 12 cylinder engines, or those used within a racing environment. A typical V12 Formula 1 engine may have two spark plugs and fuel injectors per cylinder and rev up to 20,000 RPM. The computing power needed to perform the calculations in real time is phenomenal!

The MC68332 has a CPU32 processor (MC68020 based) running at 16 MHz and a timer processor unit which has 16 channels controlled by a RISC-like processor to perform virtually any timing function. The resolution is down to 250 nanoseconds with an external clock source or 500 nanoseconds with an internal one.

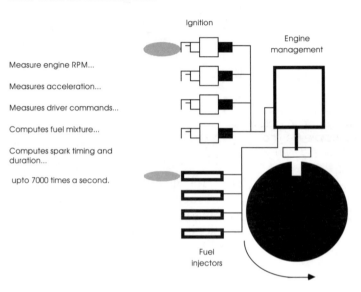

Ignition

Engine management

Measure engine RPM...

Measures acceleration...

Measures driver commands...

Computes fuel mixture...

Computes spark timing and duration...

upto 7000 times a second.

Fuel injectors

A typical engine management system

The timer processor can perform the following timer algorithms on any of the 16 channels without placing any overhead on the CPU32:

- Discrete input/output
- Pulse width modulation
- Input capture and transition
- Period measurement with additional transition detection
- Period measurement with missing transition detection
- Position synchronised pulse generation
- Stepper motor control
- Output match
- Period/pulse width accumulator

A queued serial channel and 2 Kbytes of power down static RAM are also onchip and again, for many applications, all that is required to complete a working system is an external program EPROM and a clock.

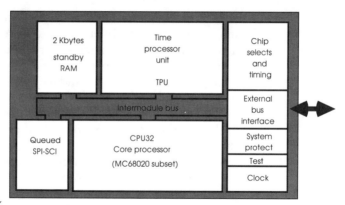

*The MC68332 block
diagram*

The advantages of an integrated processor are not restricted to the automotive industry. The MC68302 uses a 16 MHz MC68000 processor core with power down modes and either an 8 or 16 bit external bus. The system interface block contains 1,152 bytes of dual port RAM, 28 pins of parallel I/O, an interrupt controller and a DMA device, as well as the standard glue logic. The communications processor is a RISC machine that controls three multiprotocol channels, each with their own pair of DMA channels.

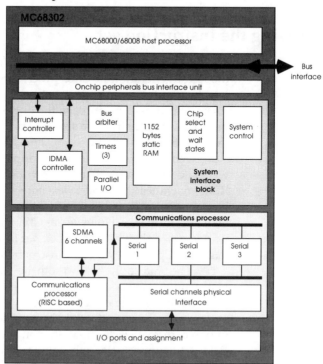

*The MC68302
processor*

The processor takes buffer structures from either the internal or external RAM and takes care of the day-to-day activities of the serial channels. It programs the DMA channel to transfer the data, performs the character and address comparisons and cyclic redundancy check (CRC) generation and checking.

The processor has sufficient processing power to cope with a combined data rate of 2 Mbits per second across the three channels. Assuming an 8 bit character and a single interrupt to poll all three channels, the processor is handling the equivalent of an interrupt every 12 microseconds. In addition, it is performing all the data framing etc. While this is going on, the onchip M68000 is free to perform other functions, like the higher layers of X.25 or other OSI protocols. The three communications channels support BISYNC, DDCMP, V.110 and HDLC synchronous modes and standard UART functions. This part essentially combines a whole single board computer bar memory and is applicable to any such application. One interesting offshoot is its use as an intelligent peripheral for both RISC and CISC systems, where the MC68302 performs all the low level I/O functions and communicates via shared memory to the main processor.

Improving the instruction set

The RISC camp has recently been further challenged by the latest generation of CISC processors, like the MC68040, which executes complex instructions in a single clock cycle. As CISC processors approach the one per clock figure, their more powerful instruction set can actually deliver more work than a RISC counterpart.

When RISC machines first appeared, CISC machines were taking about 6 to 10 clocks per instruction. This allowed RISC machines the time to execute a sequence of simpler instructions with a single clock per instruction and still offer better performance. However, as CISC processors have steadily decreased their average clock per instruction figures, this has placed pressure on RISC instruction sets.

The philosophy behind both the MC88100 and PowerPC instruction sets was simple. If a complex instruction could be executed in a single cycle and was shown to be beneficial to system performance, it would be included. This has resulted in two extremely rich sets with many addressing modes, bit field data support, floating point operations, etc. It is more accurate to describe their instruction sets as optimised, rather than reduced.

The advantages of a powerful instruction set are many. It reduces the code expansion experienced when moving from a complex to a reduced instruction set and, by reducing the number of instructions, increases the amount of work that can be performed. This may appear to be the embracing the CISC philosophy of making instructions more complex to do more work but should not be interpreted as such. Careful upgrading of RISC instruction sets, where single cycle execution and the architectural integrity can be maintained, will undoubtedly improve performance.

Upgrading instruction sets from 32 bits to 64 bits

Upgrades may not be as simple to implement as with CISC processors like the M68000 family. The difficulty concerns the fixed op code length needed by RISC machines to allow single-cycle instruction fetches and simplify the instruction decode. The MC68020 could supplement the M68000 instruction set by using operands to extend the effective op code size from two bytes to a maximum of ten. With this size, there is little difficulty in allocating bits to encode new operations. This variable length is a potential obstacle to performance improvements in that it is difficult to predict how many memory fetches are needed to fetch complete or multiple op codes in a single cycle without assuming the maximum length. The hardware difficulties of fetching 8 x 10 byte instructions are considerable.

With a 32 bit fixed op code, the size favoured by nearly all today´s RISC processors, the majority of coding bits are taken by the triadic register operands. Each instruction usually has two sources and a destination to be specified and, with a register file size of 32, this takes 15 bits, leaving 17 bits for the operation and addressing modes to be encoded. As the register file increases, each addition requires three times as many bits to encode it. With a file of over 128 registers, there are only 8 bits left to encode the instruction and any addressing mode. In these cases, the power of the instruction set is very poor and inefficient. Without destroying binary compatibility or lengthening op code size, it is difficult to see how these instruction sets can be improved. The exception to this is the movement from a 32 bit world to a 64 bit environment.

The PowerPC architecture has been defined to support both a 32 bit and a 64 bit architecture. This has been done through defining the register model as 64 bit structures but then truncating them to to a 32 bit value for the smaller implementations. This is true for all of them, including the control and status registers where the control data is stored in the 32 bit part of the register so that the information is compatible and consistent with either the 32 bit or 64 bit implementation.

The instruction encoding also remains consistent through the use of a single bit to indicate whether the instruction is working with either 32 bit or 64 bit data. In this way, software for 32 bit implementations can be recognised and executed on 64 bit machines.

Summary

It is clear that RISC technology such as PowerPC now provides the best microprocessor performance. While superscalar CISC processors can provide good levels of performance, their performance curve is fading away against that offered by the superscalar RISC processors. The reason is that their architectures have started to restrain the performance or require so much silicon real estate that it is not practical or profitable to manufacture.

With the RISC levels of performance coupled with emulation technologies, PC and workstation software can be migrated and thus the other reason for staying with CISC-based PCs and workstations is becoming weaker.

CISC is not going to disappear: it still offers very price perfomant solutions and with the expanding MC68300 family, provides excellent solutions for the embedded processor market. While CISC is no longer the most performant solution and has been replaced by RISC for sheer processing power, it will still remain the most used processor architecture for many years to come.

DSP design is following both routes: it is both improving performance and levels of integration but it is coming under fire from two areas. Microcontrollers with additional support for DSP — usually a multiply-accumulate instruction — can provide low cost and low performance alternatives to DSP. Fast RISC processors, like the PowerPC architecture, are offering the levels of performance that enable them to perform many DSP tasks that were previously beyond them. It should be remembered that DSP architectures have a lot of enhancements especially for algorithm support and so the simple comparison of MIPs or mutiply-accumulate instructions can be misleading but the differences are becoming less clear.

There is, however, one clear strategy that is emerging: processor architectures are borrowing architectural solutions from each other to solve similar problems. As a result, the architectures appear to have merged. What is even more clear is that to successfully understand the distinctions requires an understanding of RISC, CISC and DSP architectures and not just one.

13   Selecting a micro-processor architecture

It will have become apparent that there is not a single processor architecture which is the panacea to every application and, however much manufacturers attempt to declare that a processor is meant for a particular application or market, it is the ultimate use of the product which determines its suitability. This choice is made by the end user and not the supplier. Every processor has its strengths and weaknesses, its obvious and hidden applications and it can be very difficult to determine exactly the ranking the various attributes should have and their appropriate weighting. This chapter does not predetermine or dictate how processors should be used but suggests the criteria which should be considered in making such a choice.

Meeting performance needs

Until recently, the most dominant factor in choosing a processor architecture was whether it met the performance needs or not. This greatly simplified the selection process and led to a performance drive in product offerings. The manufacturer with the fastest processor generally had a big edge over any competitor. However, with the diversification of architectures available, which overlap each other in terms of performance, a far greater understanding of other factors, such as system cost and upgrading, starts to come into play. With a choice of processors offering the needed throughput, these other factors become very critical.

Choice of platforms

With the shortening design cycle and the higher competitive nature of many electronics industries, the planning of mid-life upgrades and new products becomes extremely critical. One market strategy may be to remain at the top end of the performance spectrum, in which case the overriding need for maximum throughput is the main consideration. In the case of a general purpose design, the fastest generation of RISC processors, like the PowerPC architecture will always have the performance edge.

An alternative plan may be to offer a wide portfolio of systems running common software, while providing differing performances at differing costs. In this case, the M68000 family is an ideal option. While the cost of a minimum PowerPC system will come down, it still is far higher than a lower performing M68000 or M68300 based system. While the fastest M68000 may be about 18 months behind its M88000 counterparts, the performance increases, coupled with the wide choice of platforms, may tip the balance.

In the DSP arena, code compatibility between the DSP56000 and the DSP96000 can provide another choice of platforms with different speed variations.

Whatever the choice, the system should be designed for easy upgradeability, to meet any competitive threat, improve profit margins or simply increase market share.

Anticipating future needs

The preoccupation with increasing processor throughput overshadows the other revolution that is starting within the microprocessor industry — the advent of highly integrated processors. Many applications do not need more processing throughput but require instead a higher integration level to reduce system costs and improve overall system performance. With this aspect in mind, the obvious choice is the M68000 family, with its M68300 integrated processors, or the DSP56000 with its combination of peripherals and powerful processor. The PowerPC family will undoubtedly increase the integration that it offers and support both processing pwer improvements as well as integration.

In some cases, it may be worth increasing some system costs to make an architectural switch earlier than expected, thus allowing plenty of time to become familiar with new ideas and to make the changeover in more controlled conditions, rather than under a 'make or break' regime.

Software support

Many analysts predict that software will become the dominant cost of any development involving processors and the issue of software support has quite rightly become extremely relevant. Software needs range from assemblers, compilers, debuggers and operating systems to applications and frequently the cost is underestimated.

For established architectures like the M68000, PowerPC and DSP56000 families, there is a wealth of assemblers and language support available on a multitude of hosts, ranging from simple home computers to sophisticated multiuser systems. In these situations, a selection process to determine the best software environment is necessary. The situation is similar to that of buying a hi-fi system where all the various configurations will play records and tapes, yet they offer different quality and features. In many cases, the choice may be reduced to the simple question of how much money is available.

One area that should not be forgotten is the cost of porting software to a particular hardware platform. Any system will generally be different to everybody else´s and will require even basic software, such as monitors and run time libraries, to be converted. While these ports are frequently simple to perform, they can be time consuming and costly, especially if things go wrong. Fortunately, most software companies providing these tools offer a large amount of customisation to cater for such developments — but this is not often the case for applications. Each application is usually ported to a particular set of hardware and operating system. This creates large porting revenues and is frequently a barrier for many.

This situation has developed over time and it has only been recently recognised that open standards are the key to the future. For all the criticism levelled at the IBM PC hardware, the one thing it has achieved is to awaken everyone to the advantages of shrink wrapped software. For new architectures, such as the M88000, it has been possible from day one to develop a binary compatibility standard (BCS) which allows shrink wrap software without having to enforce a 'hardware clone' on any hardware manufacturer. Cloning restricts the architectural variation that can differentiate systems. This work is now being applied to the PowerPC architecture where the PowerOpen organisation is developing a similar environment for the PowerPC architecture.

The provision of software is essential for the success of a new processor family and, rather than seen as an afterthought, it is now considered as part of the initial design program.

Development support

In the early days, the answer to this problem was simple — the manufacturer was the only source of such equipment. With the advent of cheap workstations and personal computers, this monopoly has rapidly disappeared and been replaced by a more open approach.

Development support includes providing the compiler tools and debugging support, emulation to give a window into the target environment and occasionally, some analyser tools. The trend of moving towards system simulation using behavioural models, as described in Chapter 10, is rapidly increasing. Again, the three Motorola processors are well provided for. With the M68000 family, there are many third-party products with which instead of asking what support is available, the question of 'how much do you want to spend' is raised! One aspect that does remain is the undeniable fact that the more costly and complex the system, the more sophisticated and expensive the tools will be. The standard and traditional method of system debugging was to use an emulator but this approach has been augmented by other cheaper methods.

The cheapest support is usually provided by an EVM (evaluation module), a simple board with a processor, memory, resident debugger and I/O interfaces. In many cases the board has EPROM programming facilities. Initially developed for 8 bit microcontrollers, the approach has now been brought into the 16 and 32 bit arenas. EVMs frequently come with assembler and compiler software to run on a PC or Apple Macintosh. The aim is to provide a very low cost environment, where software can be written, developed and tested and simple hardware interfaces built. Software is developed on the PC and downloaded, via a serial line to the board, where it can be run and debugged. However, they do not provide true emulation capability — i.e. the ability to replace a processor within a target system.

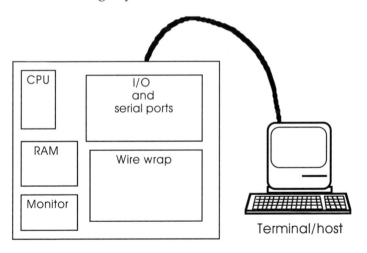

Terminal/host

An EVM system

A variation on the EVM idea is the ADS (advanced development system), where the 'EVM' is expanded by the addition of an interface card to the host and an processor probe. These systems are very simple, low cost emulators and can be used to debug a separate target system. Their facilities are not sophisticated, although they can be augmented by external logic analysers if necessary. For many simple projects, they are more than adequate.

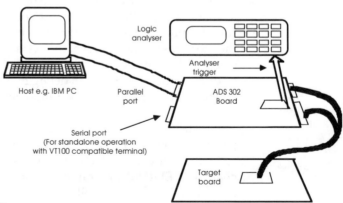

The ADS board for the
MC68302

The full blown emulator has evolved into a full blown and sophisticated piece of instrumentation. The software tools that are available allow simulation and primary testing to be performed without any target system and frequently the actual of use of an emulator is transparent to the engineer. Such systems have trace buffers, event qualification facilities and the ability to replace or overlay target memory with emulation memory within the emulator itself. Breakpoints can be set when instructions are executed or if particular signal states or addresses appear. Many such systems can actually provide multiprocessor support.

It is possible to buy an M68000 assembler, C compiler and a simple evaluation board for about £500. An emulator costs around £2,500. Similarly, a DSP56000 simulator, assembler and C compiler with a development board capable of emulation is about £3,000. System support for an MC68040 or PowerPC processor is currently far more than this! The reason is simple — the faster clock speeds and hardware require even faster hardware to gain the timing advantages necessary to provide the emulation.

Such instrumentation is therefore at least as expensive as the systems they are helping to develop. An emulator has a processor, memory, software, etc. The software tools are state-of-the-art and the many man years of development have to be paid for. In other words, development system costs frequently reflect overall system costs and sophistication.

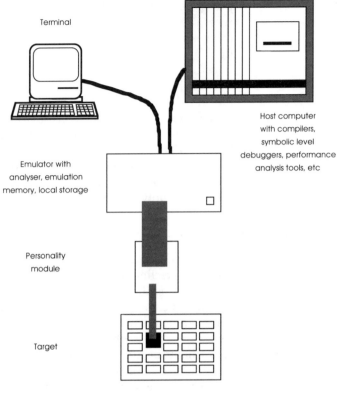

Terminal

Host computer
with compilers,
symbolic level
debuggers, performance
analysis tools, etc

Emulator with
analyser, emulation
memory, local storage

Personality
module

Target

*An integrated
emulator/analyser
with software devel-
opment tools*

Standards

Standards are becoming more important to control the costs of software development and as higher system connectivity is demanded. Standards may define communication links, quality, software conformance, etc. Apart from the standards involved with shrinkwrap software, the need for military specification components may be another factor, either for defence projects or because of hostile environments.

Again for the M68000, M88000 and DSP56000, there are well defined military programs in place. Typically, military spec parts conform to one or more of the established standards, such as 883-C or CECC. These standards define the conformance testing and associated documentation that each part has to go through before it can be supplied with its certificate of conformance and used in the end application. There are different levels of screening, depending on whether the parts have been simply processed using the standards or the appropriate institutes have actually qualified the process(es) involved. In many cases, it is the tracability in cases of failure that provide the guarantee. While screening can ensure the outgoing quality is good, it does not prevent field failures and, where a failure occurs, the product must be traced so that others of the batch can be replaced or checked.

When a new commercial part becomes available, it takes about 12 months for the first hi-rel versions to appear. Generally, they are slower than the current commercial parts, usually due to their higher operating temperature specification : − 55 to 125 degrees Celcius, compared with 0 to 70 degrees for commercial parts. Electrical timings are frequently the same, although this is not always be the case. Packages are almost certainly ceramic.

Built-in obsolescence

Is a design limited by the production technology or by the silicon technology? For many designs where production technology is the limitation, there is a great danger of the product becoming obsolete as soon as it comes to market. Semiconductors will continue to go faster and this frequently forces the pace within the marketplace. However, this increase has to be followed by a matched increase in production technologies. If a design is using older methods but stretched to the limit, the cost of changing in mid-life is far higher. The cost and difficulty of changing a circuit board may be far greater than the additional costs of faster processors and memory and therefore the upgrade path is effectively stopped, potentially making the board obsolete. The market advantage of having the highest performance system can be lost or devalued through the additional upgrade costs.

It makes sense, for a lot of systems, to design either for the highest available speed or different processor options so the design can be changed to give other advantages, such as higher speeds or a change to cheaper alternative parts.

Market changes

In the end, very few processor designs are performed simply for academic or non-commercial reason. For the majority, commercial success is as important as the engineering goals. It is now important that products are designed to provide mid-life upgrades, so a response can be made to any other competitive threat.

There are various ways in which such upgrades can be designed in. The obvious way is to provide differing processor speed and memory options within a design. This is frequently the reasoning behind the provision of a standard bus interface, like VMEbus or Multibus, which allows standard off-the-shelf products to be added with the minimum of fuss and design effort. Software packages which offer different performance levels are another method of achieving the same goal. By using improved compilers or by the removal of software delay loops, the system can be upgraded without massive costs.

Another alternative is to use the processor developments within a family to run the same system software on a range of platforms. The M68000 family can provide highly integrated low cost platforms through to super computers, allowing hardware products to be developed for various market segments, yet using a common software base.

Considering all the options

Any architectural decision must be made by considering all the relevant facts. The only way to understand the various aspects of a system and how they relate to a processor architecture is by understanding the relationships and trade-offs that are embodied within a design or application and the processor itself. The weighting given to any individual aspect must reflect the importance or necessity of that function and this, again, can only come from understanding the total system. In the final reckoning, it is the user that will make that decision.

A   'Lies, damn lies and benchmarks'

with apologies to Benjamin Disraeli

When evaluating or investigating processors, it is almost impossible to ignore the many benchmarks and counter-claims that are put forward to defend or establish which product offers the most 'bangs per buck'. In reality, the only accurate benchmark is the application running on the final end system performing real tasks. However, this is often impractical and benchmarks have to be used to give an indication of the achievable performance that can be obtained from a system or architecture. This appendix examines many of the 'tricks' used by some to enhance performance figures.

MIPS, MOPS and MFLOPS

The MIPS measurement, which gauges performance by the number of millions of instructions that are executed per second, has become a firm favourite with the media for comparing processors. The larger the number, the higher the performance. MIPS actually stands for 'meaningless indication of the performance of software', 'meaningless indication of the performance of silicon' or 'meaningless indication of the performance given by salesmen'. Take your pick depending on your background.

The problem with the MIPS measurement is that it gives no reference to what instructions are being performed and their relevant power. A 21 MIPS machine may appear to be more powerful than a 14 MIPS system but if the former takes three instructions to perform a given task while the latter can do it in one, the 14 MIPS machine will give 14 million iterations of the task to the 21 MIPS processor´s 7 million. Such a system may have great MIPS figures but is actually very inefficient in doing anything.

A variation of MIPS is the VAX MIP. Here, performance is normalised to a DEC VAX 11/780 system, where performance is rated at 1 VAX MIP. This attempts to provide some measure of the task being performed. However, the 'VAX' descriptor is often missed out and VAX MIPS compared with native instruction MIPS to greatly enhance the figures. Another trick is to normalise the performance to a VAX using a particular function in which the VAX performs very badly. This then devalues the VAX MIP rating, allowing the figures to be distorted.

The MOPS (million operations per second) is even more vague in describing what the actual operation is. Its actual meaning is 'meaningless and optimistic possibility study'. It is frequently used to enhance MIPS figures by giving even larger MOPS numbers. The argument is simple. If the system is rated at 30 MIPS but is working on 10 instructions simultaneously, it is achieving 300 MOPS! What is forgotten, is that the machine has only one pipeline and therefore can only fetch one instruction per clock. The pipeline may be 10 stages long, and therefore 10 instructions are being processed. However, common sense dictates that this is not the same as executing 10 instructions simultaneously. Actual work measurements would be typical of a 30 MIPS not a 300 MIPS machine.

Another way of looking at it is to examine an operation. An ECL NAND gate can perform a logical NAND operation with a propagation delay of 1 nanosecond, allowing 1,000,000,000 logical NAND operations per second. If there were four gates in the package, the combined throughput would be 4,000 MOPS! If this is taken to an even further extreme, every transistor in a microprocessor is capable of performing an operation. Imagine the MOPS figures that could be achieved by taking the typical switching times and multiplying by the total number. Again, this system of measurement becomes very nonsensical.

The MFLOP or million of floating point operations per second is a combination of MIPS and MOPS applied to floating point arithmetic. It usually refers to the number of instructions using floating point data that can be executed per second, although it is sometimes interpreted as operations (whatever they may be!). It suffers from the same disadvantages as MIPS and MOPS.

Unfortunately, these measurements never seem to die.

Dhrystones and Whetstones

The Dhrystone is a benchmark that has been developed for measuring integer processing power but it has been the subject of many 'dirty tricks' which have devalued it as a measurement criteria. The first version, v1.0, was written in C and intended to run on commercially available systems using an operating system such as UNIX. The intention was to provide a test of system and compiler, allowing a user to have a reasonable yardstick with which to measure system performance. The results were distributed on UNIX networks and results from 'hot boxes' or systems coded in assembler or running with no operating system were not accepted. Unfortunately, Dhrystone figures from such systems started to appear.

Some of the figures also appeared to be unbelievable. Examination of the code demonstrated one of the disadvantages — the actual benchmark does nothing except move data and compare it. Nothing is output and so some of the compiler optimizers took advantage of this, reducing the whole benchmark to a loop with a NOP in the middle. Of course the figures from such versions offered phenomenal performance. Version 1.1 fixed a bug with the previous code but was still subject to optimiser abuse. While reduction to a NOP is easy to track, there were still other tricks to be played.

One of the functions the Dhrystone benchmark performs is a string copy and comparison. Many benchmarks used a special handcoded version of these library calls to improve performance. This is not in itself despicable, providing these routines are made available to the system user. Other tricks include taking the loops and unrolling them into straightline code, removing the loop control and branching overheads. The drawback with this is increased code size. If such techniques were applied to normal applications, the resultant code expansion would be prohibitive.

Version 2.1 is now generally accepted as a better version of the benchmark and the rules prohibit loop unrolling and other disreputable practices. Even so, there are rumours of special 'benchmarking' compilers which scan the source for strings like 'Dhrystone' and automatically generate hand coded routines. In some ways, this is similar to the qualification versions of Formula 1 racing cars, which often have different engines, tyres and chassis from their actual racing counterparts. Yes, more performance may be gained — but it bears little relationship to sustained and actual delivered power.

The Whetstone is a floating point performance benchmark originally written in Fortran. It suffers from many of the disadvantages of the Dhrystone but is frequently paired with it to give a better measurement of integer and floating point performance.

Hardware effects

Hardware design is open to manipulation to increase benchmarking performance. One trick frequently used is the creation of 'hot boxes'. Such systems often have small amounts of fast static RAM and run the processor as fast as possible. Some systems even contravened the published timings but, providing they functioned sufficiently to execute the benchmark, this was fine. The problem was that the resultant performance bore little resemblance to that commercially available. A variation of this approach is to distort the comparisons by using different clock speeds. Performance figures for a 50 MHz processor or system when only 16 MHz parts are available are unacceptable. It creates an 'apples and oranges' comparison, where data is distorted and loses its validity.

The appearance of cache memory systems can further confuse the matter. Many benchmarks are small enough to fit within a cache and therefore execute with the best performance. In the real world, other applications or larger routines would be running, the cache would fill up and cause misses. Each miss can impact performance by coupling the processor direct to slow memory and, again, the benchmarked performance is not achieved by the actual system. The use of copyback or write-through schemes can also cause problems.

Optimising the system

It is very easy to obtain a five times performance increase in a general UNIX benchmark which uses standard I/O calls to disk etc. In a normal system, such communication to peripherals limits the obtainable performance. There are system variables which can be changed to allow larger buffers, disk caches and page swap areas, forcing many accesses to memory rather than peripherals. One other trick is to initialise the hard disk and freshly install the operating system. This prevents file and sector fragmentation, which can degrade system performance by about 20%. Again, the benchmark system will offer the edge over a normal machine.

These manipulations may be valid — providing they can be implemented on the end-user system.

Another trick is to remove the operating system completely and call directly to a monitor or debugger to provide the system I/O functions. Again, the benchmarked performance is far greater than practice.

Another practice is that of extrapolation. Benchmarks requiring iterative loops are simply executed by measuring the time for a single loop and extrapolating for all the iterations. This can neatly hide any problems with caches, page swapping, virtual memory, etc.

Some guidelines

Anyone entering the field of benchmarks and system performance is entering a potentially gladiatorial combat with only his wits and agility to compete with. There are three guidelines that should be considered :

•   **Establish exactly under what conditions the benchmark was run.**

Determine which compiler, processor clock speed, memory configuration and speed, operating system and system configuration, etc. were used to perform the benchmark. This information should be relatively easy to obtain, however, it is important to determine other information concerning system tuning parameters. In a UNIX environment, it is relatively easy to dramatically improve performance by re-allocating memory and disk space to paging areas, disc caches etc. With a 2 Mbyte disk cache, disk I/O performance is dramatically improved - by a factor of four or five.

•   **Examine the code and check if it has been overoptimized.**

Loop unrolling, total dead code optimisation (replacing everything with a NOP) and inlining can easily be checked. If the binary code is not available, be very suspicious! Again, it is important to determine how the optimisation has been achieved. A handcoded routine may be applicable to a real world environment — providing it is a small part of the overall system code and is frequently used. With large applications, it is nonsensical to expect the improvements offered by handcoding right across the whole application.

•   **Apply common sense.**

Obviously, a MC68000 is not going to achieve the same performance as an MC68040 or a PowerPC. Any claims where there appears to be a lack of the appropriate level of horsepower to achieve the performance figures quoted should be investigated thoroughly. It may be that a particular part of a computer system performance is being analysed and taking this in isolation distorts the overall system performance figures.

This is also true of multiprocessor and parallel processor systems. Some benchmark figures for multiprocessor systems are obtained by taking the figure for a single processor and multiplying it by the number of processors within the system. This approach is rarely valid. Such systems frequently compete for memory, mass storage and other unique system resources. The simple scaling of performance also fails to take into account any interprocessor communication overhead.

SPEC benchmarks

The SPEC organisation is a group of computer manufacturers (DEC, HP-Apollo, MIPS, Motorola, etc.) which have developed a set of benchmarks and specified the conditions used to run them in an attempt to provide a consistent environment for performance evaluation. These benchmarks include both processor-intensive tasks and user applications to mimic a real-world user environment. The initial benchmarks allowed some system tuning but this has now been removed.

These benchmarks are reasonable but, like all benchmarks, they are flawed. While providing a consistent approach, their results are based on available computer systems and do not, therefore, reflect the potential performance offered by a particular microprocessor architecture or the performance obtainable within a real-time or embedded control application. System A may be slower than System B simply because the design trade-offs were selected for good price performance rather than out and out performance. The benchmarks have been derived from a UNIX environment and their relevance to real-time applications is questionable. Having said all that, most high performance microprocessors are now quoting SPEC marks but some of them are not even run on hardware but have been simulated. How close these simulation results are to a real system design is not clear.

In the end, the best benchmark is your software, running on your hardware and performing your application!

B   Alternative micro-processor architectures

While most of the architecture examples within this book have been based on Motorola's processors, this appendix restores the balance and gives an overview of some of the other processor architectures that the reader may come across.

The overviews briefly describe the register set, internal organisation and instruction sets. For further detail on their architectural features which are frequently common in principle and understanding, refer back to the main text by using the contents or index.

INTEL 80286 microprocessor

The Intel 80286 was the successor to the 8086 and 8088 procesors and offered a larger addressing space while still preserving compatibility with its predecessors.

Architecture

The 80286 has two modes of operation known as real mode and protected mode: real mode describes its emulation of the 8086/8088 processor including limiting its external address bus to 20 bits to mimic the 8086/8088 1 Mbyte address space. In its real mode, the 80286 adds some additional registers to allow access to its larger 16 Mbyte external address space, while still preserving its compatibility with the 8086 and 8088 processors.

The register set comprises of four general purpose 16 bit registers (AX, BX, CX and DX) and four segment address registers (CS, DS, SS and ES) and a 16-bit program counter. The general purpose registers — AX, BX,CX, and DX — can be accessed as two 8 bit registers by changing the X suffix to either H or L. In this way, each half of register AX can be accessed as AH or AL and so on for the other three registers. These register form a set that is the same as that of an 8086. However, when the processor is switched into its protected mode, the register set is expanded and includes two index registers (DI and SI) and a base pointer register. These additions allow the 80286 to support a simple virtual memory scheme.

Within the IBM PC environment, the 8086 and 8088 processors can access beyond the 1 Mbyte address space by using paging and special hardware to simulate the missing address lines. This additional memory is known as expanded.

Accumulator	AX
Base register	BX
Counter reg	CX
Data register	DX
Source index	SI
Destination index	DI
Stack pointer	SP
Base pointer	BP
Code segment	CS
Data segment	DS
Stack segment	SS
Extra segment	ES
Instruction pointer	IP
Status flags	FL

15   0

Intel 80286 processor register set

Interrupt facilities

The 80286 can handle 256 different exceptions and the vectors for these are held in a vector table. The vector table's construction is different depending on the procesor's operating mode. In the real mode, each vector table consists of two 16 bit words that contain the interupt pointer and code segment address so that the associated interrupt routine can be located and executed. In the protected mode of operation each entry is eight bytes long.

Vector	Function
0	Divide error
1	Debug exception
2	Non-masked interrupt NMI
3	One byte interrupt INT
4	Interrupt on overflow INTO
S	Array bounds check BOUND
6	Invalid opcode
7	Device not available
8	Double fault
9	Coprocessor segment overrun
10	Invalid TSS
11	Segment not present

Vector	Function
12	Stack fault
13	General protection fault
14	Page fault
15	Reserved
16	Coprocessor error
17-32	Reserved
33-255	INT n trap instructions

Instruction set

The instruction set for the 80286 follows the same pattern as that for the Intel 8086 and programs written for the 8086 are compatible with the 80286 processor.

Opcode	Instruction
AAA	ASCII adjust for add
AAD	ASCII adjust for divide
AAM	ASCII adjust for multiply
AAS	ASCII adjust for subtract
ADC	Add with carry
ADD	Add
AND	Logical AND
BSF	Scan bit forward
BSR	Scan bit reverse
BT	Test bit
BTC	Test bit and complement
BTR	Test bit and reset
BTS	Test bit and set
CALL	Call subroutine
CBW	Convert byte to word
CLC	Clear carry flag
CLD	Clear direction flag
CLI	Clear interrupt enable flag
CLTS	Clear task switched flag
CMC	Complement carry flag
CMP	Compare
CMPS	Compare string byte/word
CWD	Convert word to double word
DAA	Decimal adjust for add
DAS	Decimal adjust for subtract
DEC	Decrement
DIV	Unsigned divide
ENTER	Enter procedure
HLT	Halt
IDIV	Signed integer divide
IMUL	Signed integer multiply
IN	Input from port
INC	Increment

Opcode	Instruction
INS	Input string byte/word from DX port
INT	Interrupt
INTO	Interrupt 4 if overflow set
IRET	Interrupt return
JB/JNAE	Jump on below/not above or equal
JBE/JNA	Jump on below or equal/not above
JCXZ	Jump on CX zero
JE/JZ	Jump on equal/zero
JECXZ	Jump on ECX zero
JL/JNGE	Jump on less/not greater or equal
JLE/JNG	Jump on less or equal/not greater
JMP	Unconditional jump
JNB/JAE	Jump on not below/above or equal
JNBE/JA	Jump on not below or equal/above
JNE/JNZ	Jump on not equal/not zero
JNL/JGE	Jump on not less/greater or equal
JNLE/JG	Jump not less or equal/greater
JNO	Jump on not overflow
JNP/JPO	Jump on not parity/parity odd
JNS	Jump on not sign
JO	Jump on overflow
JP/JPE	Jump on parity/parity even
JS	Jump on sign
LAHF	Load AH into flag register
LDS	Load pointer to DS
LEA	Load effective address
LEAVE	Leave procedure
LES	Load pointer to ES
LFS	Load pointer to FS
LGS	Load pointer to GS
LODS	Load string byte/word to AL/AX/EAX
LOOP	Loop CX times
LOOPZ/LOOPE	Loop while zero/equal
LOOPNZ/LOOPNE	
	Loop while not zero/not equal
LSS	Load pointer to SS
MOV	Move data
MOVS	Move string byte/word
MOVSX	Move with sign extension
MOVZX	Move with zero extension
MUL	Unsigned multiply
NEG	Negate
NOP	No operation
NOT	Logical complement
OR	Logical OR
OUT	Output to port
OUTS	Output string byte/word to DX port
POP	Pop data from stack
POPA	Pop all
POPF	Pop flags
PUSH	Push to stack

Opcode	Instruction
PUSHA	Push all
PUSHF	Push flags
RCL	Rotate left via carry
RCR	Rotate right via carry
REPECMPS	Repeated compare string (find no match)
REPNECMPS	Repeated compare string (find match)
REP INS	Repeated input string
REP LODS	Repeated load string
REP MOVS	Repeated move string
REP OUTS	Repeated output string
REPE SCAS	Repeated scan string
REPNE SCAS	Repeated scan string
REP STOS	Repeated store string
RET	Return from call
ROL	Rotate left
ROR	Rotate right
SAHF	Store flags to AH
SAL	Arithmetic shift left
SAR	Arithmetic shift right
SBB	Subtract with borrow
SCAS	Scan string byte/word
SETB/SETNAE	Set byte on below/not above or equal
SETBE/SETNA	Set byte on below or equal/not above
SETE/SETZ	Set byte on equal/zero
SETL/SETNGE	Set byte on less/not greater or equal
SETLE/SETNG	Set byte on less or equal/not greater
SETNB	Set byte on not below
SETNBE/SETA	Set byte on not below or equal/above
SETNE/SETNZ	Set byte on not equal/not zero
SETNL/SETGE	Set byte on not less/greater or equal
SETNLE/SETG	Set byte on not less or equal/greater
SETNO	Set byte on not overflow
SETNP/SETPO	Set byte on not parity/parity odd
SETNS	Set byte on not sign
SETO	Set byte on overflow
SETP/SETPE	Set byte on parity/parity even
SETS	Set byte on sign
SHL	Shift left
SHLD	Shift left double word
SHR	Shift right
SHRD	Shift right double word
STC	Set carry flag
STD	Set direction flag
STI	Set interrupt enable flag
STOS	Store string byte/word from AL/AX/EAX
SUB	Subtract
TEST	Test data
WAIT	Wait until busy pin is negated
XCHG	Exchange data
XLAT	Translate string
XOR	Logical EXCLUSIVE OR

80287 floating point support

The 80286 can also be used with the 80287 floating point coprocessor to provide acceleration for floating point calculations. If the device is not present, it is possible to emulate the floating point operations in software, but at a far lower performance.

Feature comparison

Feature	8086	8088	80286
Address bus	20 Bit	20 Bit	24 Bit
Data bus	16 Bit	8 Bit	16 Bit
FPU pesent	NO	NO	NO
Memory management	NO	NO	Yes
Cache onchip	NO	NO	NO
Branch acceleration	NO	NO	NO
TLB support	NO	NO	NO
Superscalar	NO	NO	NO
Frequency (MHz)	5,8,10	5,8,10	6,8,10,12
Average cycles/Inst.	12	12	4.9
Frequency of FPU	=CPU	=CPU	2/3 CPU
Frequency	3X	3X	2X
Address range	1 MB	1 MB	16 MB
Frequency scalibility	NO	NO	NO
Voltage	5v	5v	5v

*Intel 8086, 8088 and
80286 processors*

INTEL 80386DX

The 80386 processor was introduced in 1987 as the first 32 bit member of the family. It has 32 bit registers and both 32 bit data and address busses. It is software compatible with the previous generations through the preservation of the older register set within the 80386's newer extended register model and through a special 8086 emultaion mode where the 80386 behaves like a very fast 8086.

The processor has an onchip paging memory management unit which can be used to support multi-tasking and demand paging virtual memory schemes if required.

Architecture

The 80386 has eight general purpose 32-bit registers EAX, EBX, ECX, EDX, ESI, EDI, EBP and ESP. These general purpose registers are used for storing either data or addresses. To ensure compatibility with the earlier 8086 processor, the lower half of each register can be accessed as a 16-bit register (AX, BX, CX, DX, SI, DI, BP and SP). The AX, BX CX and DX registers can be also accessed as 8 bit registers by changing the X suffix for either H or L thus creating the 8088 registers AH,AL, BH , BL and so on.

To generate a 32 bit physical address, six segment registers (CS, SS, DS, ES, FS, GS) are used with addresses from the general registers or instruction pointer. The code segment(CS) is used with the instruction pointer to create the addresses used for instruction fetches and any stack access uses the SS register. The remaining segment registers are used for data addresses.

Each segment register has an associated descriptor register which is used to program and control the on chip memory management unit. These descriptor registers — controlled by the operating system and not normally accessible to the application programmer — hold the base address, segment limit and various attribute bits that describe the segments properties.

The 80386 can run in three different modes: the real mode, where the size of each segment is limited to 64 Kbytes, just like the 8088 and 8086. A protected mode where the largest segment size is increased to 4 Gbytes and a special version of the protected mode that creates multiple virtual 8086 processor environments.

The 32 bit flag register contains the normal carry zero, auxiliary carry, parity, sign and overflow flags. The resume flag is used with the trap 1 flag during debug operations to stop and start the processor. The remaining flags are used for system control to select virtual mode, nested task operation and input/output privilege level.

	31	15	0	31	15	0
ESI		AH AX AL	ESI		SI	
EDI		BH BX BL	EDI		DI	
EBP		CH CX CL	EBP		BP	
ESP		DH DX DL	ESP		SP	
			CS			
			DS			
			ES			
			FS			
			GS			
			SS			
		EIP		IP		
		EFLAGS		FLAGS		

Intel 80386 register set

For external input and output, a separate peripheral address facility is available similar to that found on the 8086. As an alternative, memory mapping is also supported (like the M68000 family) where the peripheral is located within the main memory map.

Interrupt facilities

The 80386 has two external interrupt signals which can be used to allow external devices to interrupt the procesor. The INTR input generates a maskable interrupt while the NMI generates a non-maskable interrupt and naturally has the higher priority of the two.

During an interrupt cycle, the processor carries out two interrupt acknowledge bus cycles and reads an 8-bit vector number on D0–D7 during the second cycle. This vector number is then used to locate within the vector table the address of the corresponding interrupt service routine. The NMI interrupt is automatically assigned the vector number of 2.

Software interrupts can be generated by executing the INT *n* instruction where *n* is the vector number for the interrupt. The vector table consists of four-byte entries for each vector and starts at memory location 0 when the processor is running in the real mode. In the protected mode, each vector is eight bytes long.

Vector	Function
0	Divide error
1	Debug exception
2	Non-masked interrupt NMI
3	One byte interrupt INT
4	Interrupt on overflow INTO
S	Array bounds check BOUND
6	Invalid opcode
7	Device not available
8	Double fault
9	Coprocessor segment overrun
10	Invalid TSS
11	Segment not present
12	Stack fault
13	General protection fault
14	Page fault
15	Reserved
16	Coprocessor error
17-32	Reserved
33-255	INT n trap instructions

Instruction set

The 80386 instruction set is essentially a superset of the 8086 instruction set. The format follows the dyadic approach and uses two operands as sources with one one of them also duplicating as a destination. Arithmetic and other similar operations thus follow the A+B=B type of format (like the M68000). When the processor is operating in the real mode — like an 8086 processor — its instruction set, data types and register model is essentially restricted to a that of the 8086. In its protected mode, the full 80386 instruction set, data types and register model becomes available. Supported data types include bits, bit fields, bytes, words (16 bit), long words (32 bit) and quad words (64 bits). Data can be signed or unsigned binary, packed or unpacked BCD, character bytes and strings.

Opcode	Instruction
AAA	ASCII adjust for add
AAD	ASCII adjust for divide
AAM	ASCII adjust for multiply
AAS	ASCII adjust for subtract
ADC	Add with carry
ADD	Add
AND	Logical AND
BSF	Bit scan forward
BSR	Bit scan reverse
BT	Test bit
BTC	Test bit and complement
BTR	Test bit and reset

Opcode	Instruction
BTS	Test bit and set
CALL	Call subroutine
CBW	Convert byte to word
CDQ	Convert double to quad word
CLC	Clear carry flag
CLD	Clear direction flag
CLI	Clear interrupt enable flag
CLTS	Clear task switched flag
CMC	Complement carry flag
CMP	Compare
CMPS	Compare string byte/word
CWD	Convert word to double word
CWDE	Convert to double word extended
DAA	Decimal adjust for add
DAS	Decimal adjust for subtract
DEC	Decrement
DIV	Unsigned divide
ENTER	Enter procedure
ESC	Escape
HLT	Halt
IDIV	Signed integer divide
IMUL	Signed integer multiply
IN	Input from port
INC	Increment
INS	Input string byte/word from DX port
INT	Interrupt
INTO	Interrupt 4 if overflow set
IRET	Interrupt return
JB/JNAE	Jump on below/not above or equal
JBE/JNA	Jump on below or equal/not above
JCXZ	Jump on CX zero
JE/JZ	Jump on equal/zero
JL/JNGE	Jump on less/not greater or equal
JLE/JNG	Jump on less or equal/not greater
JMP	Unconditional jump
JNB/JAE	Jump on not below/above or equal
JNBE/JA	Jump on not below or equal/above
JNE/JNZ	Jump on not equal/not zero
JNL/JGE	Jump on not less/greater or equal
JNLE/JG	Jump not less or equal/greater
JNO	Jump on not overflow
JNP/JPO	Jump on not parity/parity odd
JNS	Jump on not sign
JO	Jump on overflow
JP/JPE	Jump on parity/parity even
JS	Jump on sign
LAHF	Load AH into flag register
LDS	Load pointer to DS
LEA	Load effective address
LEAVE	Leave procedure
LES	Load pointer to ES

Opcode	Instruction
LFS	Load pointer to FS
LGS	Load pointer to GS
LOCK	Lock bus
LODS	Load string byte/word to AL/AX/EAX
LOOP	Loop CX times
LOOPZ/LOOPE	Loop while zero/equal
LOOPNZ/LOOPNE	Loop while not zero/not equal
LSS	Load pointer to SS
MOV	Move data
MOVS	Move string byte/word
MOVSX	Move with sign extension
MOVZX	Move with zero extension
MUL	Unsigned multiply
NEG	Negate
NOP	No operation
NOT	Logical complement
OR	Logical OR
OUT	Output to port
OUTS	Output string byte/word to DX port
POP	Pop data from stack
POPA	Pop all
POPF	Pop flags
POPFD	Pop Eflags from stack
PUSH	Push to stack
PUSHA	Push all
PUSHF	Push flags
PUSHFD	Push Eflags to stack
RCL	Rotate left via carry
RCR	Rotate right via carry
REPE CMPS	Repeated compare string (find no match)
REPNE CMPS	Repeated compare string (find match)
REP INS	Repeated input string
REP LODS	Repeated load string
REP MOVS	Repeated move string
REP OUTS	Repeated output string
REPE SCAS	Repeated scan string
REPNE SCAS	Repeated scan string
REP STOS	Repeated store string
RET	Return from call
ROL	Rotate left
ROR	Rotate right
SAHF	Store flags to AH
SAL	Arithmetic shift left
SAR	Arithmetic shift right
SBB	Subtract with borrow
SCAS	Scan string byte/word
SETB/SETNAE	Set byte on below/not above or equal
SETBE/SETNA	Set byte on below or equal/not above
SETE/SETZ	Set byte on equal/zero
SETL/SETNGE	Set byte on less/not greater or equal

Opcode	Instruction
SETLE/SETNG	Set byte on less or equal/not greater
SETNB	Set byte on not below
SETNBE/SETA	Set byte on not below or equal/above
SETNE/SETNZ	Set byte on not equal/not zero
SETNL/SETGE	Set byte on not less/greater or equal
SETNLE/SETG	Set byte on not less or equal/greater
SETNO	Set byte on not overflow
SETNP/SETPO	Set byte on not parity/parity odd
SETNS	Set byte on not sign
SETO	Set byte on overllow
SETP/SETPE	Set byte on parity/parity even
SETS	Set byte on sign
SHL	Shift left
SHLD	Shift left double word
SHR	Shift right
SHRD	Shift right double word
STC	Set carry flag
STD	Set direction flag
STI	Set interrupt enable flag
STOS	Store string byte/word from AL/AX/EAX
SUB	Subtract
TEST	Test data
WAIT	Wait until busy pin is negated
XCHG	Exchange data
XLAT	Translate string
XOR	Logical EXCLUSIVE OR

In addition, there is a further group of instructions that can be used when the CPU is running in protected mode only. They provide access to the memory management and control registers. Typically, there are not available to the user programmer and are left to the operating system to use.

Opcode	Instruction
ARPL	Adjust requested privilege level
LAR	Load access rights
LGDT	Load global descriptor table
LIDT	Load interrupt descriptor table
LLDT	Load local descriptor table
LMSW	Load machine status word
LSL	Load segment limit
LTR	Load task register
SGDT	Store global descriptor table
SIDT	Store interrupt descriptor table
STR	Store task register
SLDT	Store local descriptor table
SMSW	Store machine status word
VERR	Verify segment for reading
VERW	Verify segment for writing

Addressing modes provided are:

Register direct	(Register contains operand)
Immediate	(Instruction contains data)
Displacement	(8/16 bits)
Base address	(Uses BX or BP register)
Index	(Uses DI or SI register)

80387 Floating point coprocessor

The 80386 can also be used with the 80387 floating point coprocessor to provide acceleration for floating point calculations. If the device is not present, it is possible to emulate the floating point operations in software, but at a far lower performance.

Feature comparison

There is a derivative of the 80386DX called the 80386SX which provides a lower cost device while retaining the same architecture. To reduce the cost, it uses an external 16 bit data bus and a 24 bit memory bus. The SX device is not pin compatible with the DX device.

In addition, Intel have produced an 80386SL device for portable PCs which incorporates a power control module that provides support for efficient power conservation.

Although Intel designed the 80386 series, the processor has been successfully cloned by other manufacturers (both technically and legally) such as AMD and Cyrix. Their versions are available at far higher clock speeds that the Intel originals and many PCs are now using them.

Feature	i386SX	i386DX	i386SL
Address bus	24 Bit	32 Bit	24 Bit
Data bus	16 Bit	32 Bit	16 Bit
FPU present	No	No	No
Memory management	Yes	Yes	Yes
Cache onchip	No	No	Control
Branch acceleration	NO	NO	NO
TLB support	NO	NO	NO
Superscalar	NO	NO	NO
Frequency (MHz)	16,20,25,33	16,20,25,33	16,20,25"
Avg. cycles/inst.	4.9	4.9	<4.9
Frequency of FPU	=CPU	=CPU	=CPU
Address range	16 Mbytes	4 Gibytes	16 Mbytes
Freq. scalability	No	No	No
Transistors	275,000	275,000	855,000
Voltage	5v	5v	3v or 5v
System management	No	No	Yes

INTEL 80486

The Intel 80486 processor is essentially an enhanced 80386. It has a similar instruction set and register model but to dismiss it as simply a go-faster 80386 would be ignoring the other features that it uses to improve performance. Like the MC68040, it is a CISC processor that can execute instructions in a single cycle. This is done by pipelining the instruction flow so that adress calculations and so on are performed as the instruction proceeds down the line. Although the pipeline may take several cycles, an instruction can potentially be started and completed on every clock edge, thus achieving the single cycle performance.

To provide instruction and data to the pipeline, the 80486 has an internal unified cache to contain both data and instructions. This removes the dependency of the processor on faster external memory to maintain sufficient data flow to allow the processor to continue executing instead of stalling.

The 80486 also integrates a 80387 compatible fast floating point unit and thus does not need an external coprocessor.

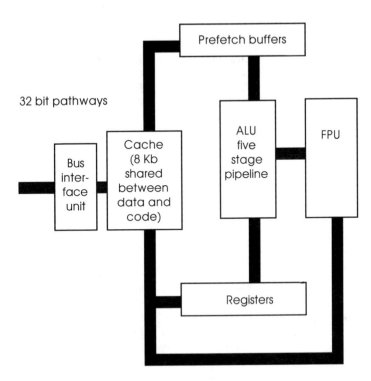

The 80486 internal architecture

Instruction set

The instruction set is essentially the same as the 80386 but there are some additional instructions available when running in protected mode to control the memory mangement and floating point units.

Memory management instructions

Opcode	Instruction
ARPL	Adjust requested privilege level
LAR	Load access rights
LGDT	Load local descriptor table
LIDT	Load interrupt descriptor table
LLDT	Load local descriptor table
LMSW	Load machine status word
LSL	Load segment limit
LTR	Load task register
SGDT	Store global descriptor table
SIDT	Store interrupt descriptor table
STR	Store task register
SLDT	Store local descriptor table
SMSW	Store machine status word
VERR	Verify segment for reading
VERW	Verify segment for writing

Floating point instructions

Opcode	Instruction
FCOM	Compare ST(0) with Real
FCOMP	FCOM and Pop
FCOMPP	FCOM and Pop twice
FICOM	Compare ST(0) with Integer
FICOMP	FICOM and Pop
FTST	Compare ST(0) with 00
FUCOM	Unordered compare ST(0) with ST(l)
FUCOMP	FUCOM and Pop
FXAM	Examine ST(0)
FLD	Real load to ST(0)
FILD	Integer load to ST(0)
FBLD	BCD load to ST(0)
FST	Store RealFrom ST(0)
FSTP	Store RealFrom ST(0) and Pop
FIST	Store IntegerFrom ST(0)
FISTP	Store Integer and Pop
FBSTP	Store BCD and Pop
FLTZ	Load 00 into ST(0)
FLD1	Load 10 Into ST(0)
FLDPI	Load Pi into ST(0)
FLDL2T	Load log2(10) to ST(0)
FLDL2E	Load log2(e) to ST(0)
FLDLG2	Load logl0(2) to ST(0)

Opcode	Instruction
FADD	Add Real to ST(0)
FADDP	FADD and Pop
FSUB	Subtract RealFrom ST(0)
FSUBP	FSUB and Pop
FSUBR	Reverse subtract Real
FSUBRP	FSUB and Pop
FMUL	Multiply Real with ST(0)
FMULP	FMUL and Pop
FDIV	Divide ST(0) by Real
FDIVP	FDIV and Pop
FDIVR	Reverse divide Real
FDIVRP	FDIVR and Pop
FIADD	Add Integer to ST(0)
FISUB	Subtract IntegerFrom ST(0)
FISUBR	Reverse subtract Integer
FIMUL	Multiply Integer with ST(0)
FIDIV	Divide ST(0) by Integer
FIDIVR	Reverse divide Integer
FSQRT	Square root
FSCALE	Scale ST(0) by ST(1)
FXTRACT	Extract components of ST(0)
FPREM	Partial remainder
FPREMI	Partial remainder IEEE
FRDNINT	Round ST(0) to Integer
FABS	Absolute value of ST(0)
FCHS	Change sign of ST(0)
FCOS	Cosine of ST(0)
FPTAN	Partial tangent of ST(0)
FPATAN	Partial ARCTAN
FSIN	Sine of ST(0)
FSINCOS	Sine and cosine of ST(0)
F2XM1	$2ST(°)-1$
FYL2X	$ST(1) \times log2(ST(O))$
FYL2XP1	$ST(1) \times log2(ST(O) + 10)$
FINIT	InitialiseFPU
FSTSW	Store status word
FLDCW	Load control word
FCLEX	Clear exceptions
FSTENV	Store environment
FLDENV	Load environment
FSAVE	Save state
FRSTOR	Restore state
FINCSTP	Increment stack pointer
FDECSTP	Decrement stack pointer
FFREE	Free ST(1)
FNOP	No operation
WAIT	Wait untilFPU ready

Intel486 SX and overdrive processors

The 80486 is available in several different versions which offer different facilities. The 486SX is like the 80386SX a stripped down version of the full DX processor with the floating point unit removed but with the normal 32 bit external data and address busses. The DX2 versions are the clock doubled versions which run the internal processor at twice the external bus speed. This allows a 50 MHz DX2 processor to work in a 25 MHz board design. This opens the way to retrospective upgrades — known as the overdrive philosphy — where a user simply replaces a 25MHz 486SX with a DX to to get floating point support or a DX2 to get the FPU and theoretically twice the performance. Such upgrades need to be carefully considered: removing devices that do not have a zero insertion force socket can be tricky at best and wreck the board at worst. Similarly, the additional heat and power dissipation has also to be taken into consideration. While some early PC designs had difficulties in these areas, the overdrive option has now become a standard PC option.

The DX2 typically gives about 1.6 to 1.8 perfomance improvement depending on the operations that are being carried out. Internal processing gains the most from the DX2 approach while memory-intensive operations are frequently limited by the external board design.

Intel have also released a DX4 version which offers internal CPU speeds of 75 and 100 Mhz.

Feature	i486DX2-40	i486DX2-50	i486DX2-66
Address bus	32 bit	32 bit	32 bit
Data bus	32 bit	32 bit	32 bit
FPU present	Yes	Yes	Yes
Memory management	Yes	Yes	Yes
Cache onchip	8K unified	8K unified	8K unified
Branch acceleration	NO	NO	NO
TLB support	NO	NO	NO
Superscalar	NO	NO	NO
Frequency (MHz)	40	50	66
Avg. cycles/inst	1.03	1.03	1.03
Frequency of FPU	=CPU	=CPU	=CPU
Upgradable	Yes	Yes	Yes
Address range	4 Gbytes	4 Gbytes	4 Gbytes
Freq. scalability	No	No	No
Transistors	1.2 Million	1.2 Million	1.2 Million
Voltage	5V and 3V	5V and 3V	5V

Feature	i486SX	i486DX	i486DX-50
Address bus	32 bit	32 bit	32 bit
Data bus	32 bit	32 bit	32 bit
FPU present	No	Yes	Yes
Memory management	Yes	Yes	Yes
Cache onchip	8K unified	8K unified	8K unified
Branch acceleration	NO	NO	NO
TLB support	NO	NO	NO
Superscalar	NO	NO	NO
Frequency (MHz)	16,20,25,33	25,33	50
Avg. cycles/inst.	1.03	1.03	1.03
Frequency of FPU	N/A	=CPU	=CPU
Upgradable	Yes	Yes	NO
Address range	4 Gbytes	4 Gbytes	4 Gbytes
Freq. scalability	NO	NO	NO
Transistors	1.2 Million	1.2 Million	1.2 Million
Voltage	5V and 3V	5V and 3V	5V
System management	NO	NO	NO

Intel PENTIUM

The PENTIUM is essentially an enhanced 80486 from a programming model. It uses virtually the same programming model and instruction set — although there are some new additions.

The most noticable enhamcement is its ability to operate as a superscalar processor and execute two instructions per clock. To do this it has incorporated many new features that were not present on the 80486.

As the internal architecture diagram shows, the device has two five-stage pipelines that allow the joint execution of two integer instructions providing that they are simple enough not to use microcode or have data dependencies. This restriction is not that great a problem as many compilers have now started to concentrate on the simpler instructions within CISN instruction sets to improve their performance.

To maintain the throughput, the unified cache that appeared on the 80486 has been dropped in favour of two separate 8 kbyte caches: one for data and one for code. These caches are fed by an external 64 bit wide burst mode type data bus. The caches also now support writeback MESI policies instead of the less efficient write-through design.

Branches are accelerated using a branch target cache and works in conjunction with the code cache and prefetch buffers.The instruction set now supports an eight byte compare and exchange instruction and a special processor identification instruction. The cache coherency support also has some new instructions to allow programmer's control of the MESI coherency policy.

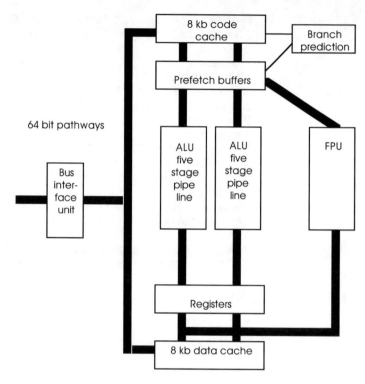

The Intel Pentium
inetrnal architecture

The Pentium has an additional control register and the system management mode register that first appeared on the 80386SL which provides intelligent power control.

Feature	Pentium
Address bus	32 bit
Data bus	64 bit
FPU present	Yes
Memory management	Yes
Cache onchip	Two 8K caches (data and code)
Branch acceleration	Yes – branch target cache
Cache coherency	MESI protocol
TLB support	Yes
Superscalar	Yes (2)
Frequency (MHz)	60, 66, 75, 100
Avg. cycle/inst.	0.5
Frequency of FPU	=CPU
Address range	4 Gigabytes
Frequency scalability	No
Transistors	3.21 million
Voltage	5 and 3V
System management	Yes

MIPS R2000/R3000

The MIPS R2000 and R3000 processors are fairly unique in the processor world in that their development was outside of a major semiconductor or computer company. The original designs were made by MIPS based in California and then licensed to various manufacturers such as LSI, NEC, Siemens and Performance Semiconductor. As a result, the processor designs have quickly spawned many derivatives.

The architecture is based on a design developed at Stanford University California. The R3000 is an enhanced version of the R2000 which can provide higher execution speed and features a different memory interface to support larger external data and instruction caches, so necessary to provide data and instructions to a RISC processor.

The R2000 and R3000 designs provide floating point support via an external coprocessor. Floating point instructions are automatically routed to the coprocessor via the external bus so that their use appears transparent to the programmer.

The main processor unit has memory management facillities, including a translation lookaside buffer, to provide 4 Gbyte virtual addressing space. However, this space is split into two to give the user mode a 2 Gbyte virtual address space.

Architecture

The main register set consists of a bank of thirty two 32 bit general purpose registers which can support either the big or little endian byte structure and can support both four, two and single byte data.

Multiplication and division operations use two dedicated 32 bit registers. Status information is provided by ten registers as shown:

Register	Description
ENTRY HI	High part of TLB entry
ENTRY LO	Low part of TLB entry
INDEX	Index pointer into TLB
RANDOM	Random pointer into TLB
STATUS	Mode, interrupt and status flags
CAUSE	Shows cause of last exception
EPC	Program counter for exceptions
CONTEXT	Context pointer
BAD VA	Most recent bad virtual address
PR ID	Processor revision code

The 32 bit external memory bus accesses the main memory via two external data caches constructed out of external SRAM. The processor provides the control signals that are necessary to control the data and instruction caches. By using external caches, the processor cost can be kept at a minimum although the external design is now more difficult especially with faster designs over 25 Mhz.

Interrupt facilities

Six hardware interrupt inputs INTR0–INTR5 are provided. These have a dual purpose: during reset they provide configuration information concerning the byte ordering and the size of the external caches. After reset, they return to their more normal operation.

Instruction set

The instruction set follows the normal RISC tenet of simplification and reduction and consists of 74 basic instructions.

Each instruction is a single 32 bit word and contains two sources and a single destination as well as the op code. Addressing modes are greatly simplified: register, register indirect or with a 16 bit immediate value. Branching is performed using a 16 bit offset from the program counter while jumps can use a 26 bit address or a register as the target address.

Branching is not hardware accelerated and like many of the techniques used within the MIPS architectures relies on sophisticated compiler technology to accelerate it. The compiler technology recognises delay slots caused by branch calculations and reorders the code so that other downstream instructions can be inserted to regain the lost performance. Needless to say, this technique must be done carefully to ensure that the re-ordering does not destroy the program's syntax or context.

Opcode	Description
ADD	Add
ADDI	Add immediate
ADDIU	Add immediate unsigned
ADDU	Add unsigned
AND	Logical AND
ANDI	Logical AND immediate
BCzF	Branch on coprocessor z false
BCzT	Branch on coprocessor z true
BEQ	Branch on equal
BGEZ	Branch on greater or zero
BGEZAL	Branch on greater or zero and link

Opcode	Description
BGTZ	Branch on greater than zero
BLEZ	Branch on less or zero
BLTZ	Branch on less than zero
BLTZAL	Branch on less than and link
BNE	Branch on not equal
BREAK	Break
COPz	Coprocessor operation
CTCz	Move control from coprocessor
CTCz	Move control to coprocessor
DIV	Divide
DIVU	Unsigned divide
J	Jump
JAL	Jump and link
JR	Jump to register
JALR	Jump and link register
LB	Load byte
LBU	Load unsigned byte
LH	Load half word
LHU	Load unsigned half word
LUI	Load upper immediate
LW	Load word
LWCz	Load word to coprocessor
LWL	Load word left
LWR	Load word right
MFC0	Move from control coprocessor
MFCz	Move from coprocessor
MFHI	Move from HI
MFLO	Move from LO
MTC0	Move to control coprocessor
MTCz	Move to coprocessor
MTHI	Move to HI
MTLO	Move to LO
MULT	Multiply
MULTU	Unsigned multiply
NOR	Logical NOR
OR	Logical OR
ORI	Logical OR immediate
RFE	Return from exception
SB	Store byte
SH	Store half word
SLL	Logic shift left
SLLV	Logic shift left variable
SRA	Arithmetic shift right
SRAV	Arithmetic shift right variable
SRL	Logic shift right
SRLV	Logic shift right variable
SLT	Set on less than
SLTI	Set on less than immediate
SLTIU	Set on less than immediate unsigned
SLTU	Set on less than unsigned
SUB	Subtract

Opcode	Description
SUBU	Subtract unsigned
SW	Store word
SWCz	Store word to coprocessor
SWL	Store word left
SWR	Store word right
SYSCALL	System call
TLBP	Probe TLB for entry match
TLBR	Read indexed TLB entry
TLBWI	Write indexed TLB entry
TLBWR	Write random TLB entry
XOR	EXCLUSIVE OR
XORI	EXCLUSIVE OR immediate

Support chips

R2010 Floating point coprocessor
R2020 Write buffers for main memory
R3010 Floating point coprocessor for R3000
R3020 Write buffers for main memory (R3000)

MIPS R4000

The R4000 took the existing R2000/R3000 processor architecture into a 64 bit environment and multiple instruction per clock arena. Its register set was expanded to 64 bits thus supporting 64 bit data. The virtual address space was also increased to give a 64 bit range. The instruction width has remained consistent with the previous processors and is still 32 bits wide.

The device is capable of executing multiple instructions per clock but it does not use a superscalar technique. Instead the execution pipeline is double pumped to enable two instructions to be processed in the time normally taken by one. To the system, it appears to have executed two instructions per clock. Again, software techniques are extensively used to isolate conflicting instructions that prevent or stall the double pumping through re-ordering. One direct result of this technique is that there are no duplication of the execution unit, thus reducing the transistor count, die size and cost.

The R4000 has also integrated onchip caches and floating point units while still preserving support for external caches using discrete SRAM.

SUN SPARC RISC processor

The SPARC (scalable processor architecture) processor is a 32 bit RISC architecture developed by Sun Microsystems for their workstations but manufactured by a number of manufacturers such as LSI, Cypress, Fujitsu, Philips and Texas Instruments.

The basic architecture follows the Berkeley model and uses register windowing to improve context switching and parameter passing. The initial designs were based on a discrete solution with separate floating point units, memory management facilities and cache memory, but later designs have integrtaed these versions. The latest versions also support superscalar operation.

Architecture

The SPARC system is based on the Berkeley RISC architecture. A large 32 bit wide register bank containing 120 registers is divided into a set of seven register windows and a set of eight registers which are globally available. Each window set containing 24 registers are split into three sections to provide eight input, eight local and eight output registers. The registers in the output section provide the information to the eight input registers in the next window. If a new window is selected during a context switch or as a procedural call, data can be passed with no overhead by placing it in the output registers of the first window. This data is then available to the procedure or next context in its input registers. In this way, the windows are linked together to form a chain where the input registers for one window have the contents of the output registers of the previous window.

To return information back to the original or calling software, the data is put into the input registers and the return executed. This moves the current window pointer back to the previous window and the returned information is now available in that window's output registers. This method is the reverse of that used to initially pass the information in the first place.

The programmer and CPU can track and control which windows are used and what to do when all windows are full, through fields in the status register.

The architecture is also interesting in that it is one of the few RISC processors that uses logical addressed caches intead of physically addressed caches.

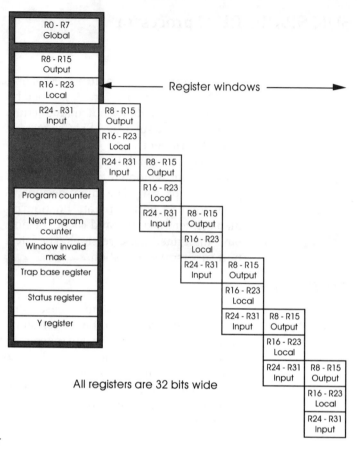

The SPARC register model

Interrupts

The SPARC processor supports 15 external interrupts which are generated using the four interrupt lines, IRL0 to IRL3. Level 15 is assigned as a non-maskable interrupt and the other 14 can be masked if required.

An external interrupt will generate an internal trap where the current and the next instructions are saved, the pipeline flushed and the processor switched into supervisor mode. The trap vector table whose location is located in the trap base register is then used to supply the address of the service routine. When the routine has completed, the REIT instruction is executed which restores the processor status and allows it to continue.

Instruction set

The instruction set comprises of of 64 instructions. All access to memory is via load and store instructions as would be expected with a RISC architecture. All other instructions operate on the register set including the currently selected window. The instruction set is also interesting in that it has a multiply step command instead of the more normal multiply command. The multiply step command allows a multiply to be synthesized.

Opcode	Description
ADD	Add
ADDX	Add with carry
AND	Logical AND
ANDN	Logical NAND
Bicc	Branch on integer condition codes
CALL	Call
CBccc	Branch on coprocessor condition codes
CPop	Coprocessor operate
FBfcc	Branch on FP condition codes
FPop	Floating point operate
IFLUSH	Flush instruction cache
JMPL	Jump and link
LD	Load word
LDC	Load coprocessor
LDCSR	Load coprocessor state
LDD	Load double word
LDDC	Load double coprocessor
LDF	Load floating point
LDDF	Load double floating point
LDFSR	Load floating point status
LDSB	Load signed byte
LDSH	Load signed half word
LDSTUB	Atomic load/store unsigned byte
LDUB	Load unsigned byte
LDUH	Load unsigned half word
MULScc	Multiply step
OR	Logical OR
ORN	Logical NOR
RDPSR	Read processor state
RDTBR	Read trap base register
RDWIM	Read window invalid mask
RDY	Read Y register
RESTORE	Restore caller's window
RETT	Return from trap
SETHI	Set high 22 bits of register
SLL	Logical shift left
SRL	Logical shift right
SRA	Arithmetic shift right
ST	Store word
STB	Store byte
STC	Store coprocessor
STC	Store double coprocessor
STCSR	Store coprocessor state
STDCQ	Store double coprocessor queue

Opcode	Description
STD	Store double word
STDF	Store double floating point
STDFQ	Store double FP queue
STF	Store floating point
STFSR	Store floating point status
STH	Store half word
SUB	Subtract
SUBX	Subtract with carry
SWAP	Swap register with memory
TADD	Tagged add
Ticc	Trap on integer condition codes
TSUBcc	Tagged subtract
UNIMP	Unimplemented instruction
WRPSR	Write processor status
WRTBR	Write trap base register
WRWIM	Write window invalid mask
WRY	Write Y register
XOR	Logical EXCLUSIVE OR
XORN	Logical EXCLUSIVE NOR

DEC Alpha

The Alpha RISC processor from DEC has achieved some notable firsts in terms of clock speed, MIPS performance and power dissipation but has had limited success in obtaining a wider adoption beyond DEC mini-computers and workstations.

It is a true 64 bit architecture with a 64 bit wide register set and support for 64 bit integers and 64 bit virtual addressing. To support the VMS operating system, it can also support 32 bit address maps as well. Internally it is a superscalar Harvard architecture which is fed by two internal caches and a 64 bit wide external memory bus. The initial architecture is designed to allow two instructions to be issued in each clock cycle but higher levels of superscalar operation are promised for the future.

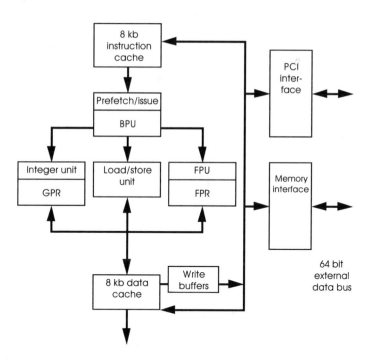

DEC Alpha internal architecture

With a total transistor count of 1.68 million devices, the 21064 processor is a complete integrated unit, including full integer and floating point execution units. These units, together with related addressing and branching units, are fully pipelined, and each is capable of launching a new operation every cycle.

In addition, the chip includes two high speed primary caches. An 8 Kbyte instruction cache provides two full 32 bit instructions per clock cycle to the instruction dispatch unit, and an 8 Kbyte data cache can provide 64 bit data access during each cycle.

The register set is   has 32 integer registers and 32 floating registers, each 64 bits and supports a load-store architecture. Integer register R31 and floating register F31 are always zero (the PowerPC and M88000 use r0).

Instructions are 32 bits wide and use the triadic format. An instruction will use source registers RA and RB and write the result in register RC. There is an extended opcode in the 11 bit function field. Integer operates can use the RB field and part of the function field to specify an 8 bit, zero-extended literal.

The most remarkable aspect of the Alpha design is the clock speed that it uses. The first parts had a 200 MHz clock which has subsequently been uprated to 300 MHz. This coupled with its superscalar operation gives very high MIPS ratings (read appendix A before placing any real interpretation on these figures). However, there is a penalty for this in that the device dissipates a lot of heat. A typical figure of 40 W has been quoted for the 300 Mhz processor.

Instruction set

The instruction set is fairly conventional but does include special support for the VMS operating system and for users migrating from the VAX environment towards Alpha. The PALcall support is a good example of this.

PALcall instructions

The privileged architecture library (PALcall) instructions specify one of a few dozen complex functions to be performed. These functions deal with interrupts and exceptions, task switching, virtual memory, and other complex operations that must be done atomically. PALcall instructions branch to a privileged library of software subroutines (using the same Alpha instruction set) that implement an operating system specific set of these complex operations.

Branch instructions

Conditional branch instructions can test a register for positive/negative or for zero/non-zero. They can also test integer registers for even/odd. Unconditional branch instructions can write a return address into a register. There is also a calculated jump instruction that branches to an arbitrary 64 bit address in a register. Branch prediction is supported.

Load/store instructions

Load and store instructions can move either 32 or 64 bit aligned quantities. Data is fetched from a linear 64 bit virtual addresses with no segmentation.

There are no 8 or 16 bit load/store instructions, but there are facilities for doing byte manipulation in registers. In addition, the processor implementation supports load and store re-ordering to improve bandwidth utilisation.

Integer operate instructions

The integer operate instructions manipulate full 64 bit values and include the usual assortment of arithmetic, compare, logical, and shift instructions. There are just three 32 bit integer operates: add, subtract, and multiply. These differ from their 64 bit counterparts only in overflow detection and in producing 32 bit results. There is no integer divide instruction.

Floating-point operate instructions

The floating point operate instructions include four complete sets of VAX and IEEE arithmetic, plus conversions between float and integer. There is no floating point square root instruction.

Index

G

H

I

J